The Montana Mathematics Enthusiast

Monograph 2

Mathematics Education and the Legacy of Zoltan Paul Dienes

The Montana Mathematics Enthusiast

Monograph 2

*Mathematics Education and the Legacy
of Zoltan Paul Dienes*

Edited by
Bharath Sriraman
The University of Montana

INFORMATION AGE PUBLISHING, INC.
Charlotte, NC • www.infoagepub.com

Library of Congress Cataloging-in-Publication Data

Mathematics education and the legacy of Zoltan Paul Dienes / edited by
Bharath Sriraman.
    p. cm. – (The Montana mathematics enthusiast monograph series in
mathematics education)
  Includes bibliographical references.
  ISBN 978-1-59311-897-6 (hardcover) – ISBN 978-1-59311-896-9 (pbk.)
1. Mathematics–Study and teaching. 2. Dienes, Zoltan P. (Zoltan Paul)
I. Dienes, Zoltan P. (Zoltan Paul) II. Sriraman, Bharath.
  QA11.2.M27778 2008
  510.71–dc22

                                                              2008003135

CONTENTS

PREFACE

THE LEGACY OF ZOLTAN PAUL DIENES

Bharath Sriraman
The University of Montana

The second monograph of The Montana Mathematics Enthusiast contains a unique collection of articles which span five generations of mathematics educators. As the title indicates the monograph is centered on the work of Zoltan Paul Dienes (1916–), a living legend in the field of mathematics education, for his pioneering work that has spanned 50 years. Trained as a mathematician in England, Zoltan became interested in the psychology of learning in the 1950s and earned a second degree in psychology. Extant histories of the field of mathematics education are often outlined in terms of its origins in the classical tradition of Felix Klein onto the structuralist agenda influenced by the Bourbaki and Dieudonné at the Royaumont seminar in post world war II France, followed by Freudenthal's reconception of mathematics education with emphasis on the humanistic element of doing mathematics (e.g., D'Ambrosio, 2003; Skovsmose, 2005). While the approaches of Klein and Dieudonné steeped in an essentialist philosophy gave way to the pragmatic approach of Freudenthal, Zoltan's approach influenced by structuralism and cognitive psychology remains unique from the point of view of developing a theory of learning which has left a lasting

Mathematics Education and the Legacy of Zoltan Paul Dienes, pages vii–x
Copyright © 2008 by Information Age Publishing

impact on the field. The books *Building up Mathematics* (Dienes, 1960) and *Thinking in Structures* (Dienes & Jeeves, 1965) influenced mathematics educators entering the field in the late 1960's and 1970's and remain classics to date. I encountered these two books in the late 1990's in graduate school when I became interested in cognition, particularly how the process of generalization could be systematically studied and understood in the context of creating new mathematics. Dienes' name is synonymous with the Multibase blocks which he invented for the teaching of place value. Among numerous other things, he is also the inventor of Algebraic materials and logic blocks, which sowed the seeds of contemporary uses of manipulative materials in instruction. Dienes' place is unique in the field of mathematics education not only because of his theory on how mathematical structures can be effectively taught from the early grades onwards using manipulatives, games, stories and dance (e.g., Dienes, 1973), but also because of his tireless attempts for over 50 years to inform school practice through his fieldwork in the UK, Italy, Australia, Brazil, Canada, Papau New Guinea and the United States. In my conversations with Richard Lesh, it was interesting to learn from him first hand the influence Zoltan had on his generation of researchers, and the foundations of Dienes' theories on the learning of mathematics on those involved in the Rational Number Project and more recently those working in the models and modeling area of research (see Lesh et al., 1987, 1989, 1992, 2003).

I had the honor of finally meeting Zoltan Dienes in Wolfville, Nova Scotia in April 2006 after many years of correspondence. The formal interview during the visit is part of this monograph (see Sriraman & Lesh, 2007). At this time, Zoltan was still actively engaged in creative and generative activities aimed at children discovering non-trivial mathematical structures. During this visit he gave me the last existing copy of a manuscript which he had written in 1995. This unpublished manuscript entitled *"Some thoughts on the dynamics of learning mathematics"* contained the "finished" form of Zoltan's theory of learning mathematics which he began developing in the 1960's. Even though an electronic copy of this manuscript was unavailable, the manuscript was in good enough condition to be scanned and reproduced in pdf format. I was particularly pleased to be able to include this entire (un-edited) manuscript in the online version of the monograph. This manuscript revealed some of the interactions of Zoltan Dienes with Jean Piaget as well as Jerome Bruner, and the esteem in which he was held by these researchers. At the age of 91, Zoltan continues to write and publish articles in numerous journals in New Zealand and the U.K. (see Sriraman & Lesh, 2007 for these citations). In this book, I am pleased to bring four articles written by Zoltan Dienes in the last six months which reveal the power of manipulatives and imaginative thinking in facilitating the discovery of beautiful mathematical structures. Dienes championed the use of collaborative group work and concrete materials, as well as goals such as democratic access to the myriad processes of mathematical thinking, long before the words "constructivism" and "equity" and "democratization" became fashionable. As Dick Lesh states, Dienes not only studied a phenomena that later cognitive scientists have come to call embodied knowledge and situated cognition—but he also emphasized the *multiple embodiment principle* whereby students need to make predictions from one structured situation to another. And, he also emphasized the fact that, when conceptual systems are partly off-loaded from the mind using a variety of interacting representational systems that every such model is at best a useful oversimplification of, both the underlying conceptual systems being expressed and the external systems that are being described or explained (Sriraman & Lesh, 2007).

The monograph includes one chapter by Lyn English, which explores the relationship between cognitive psychology and mathematics education. "Cognitive Psychology and Mathematics Education: Reflections on the past and future" contain her reflections on issues of continued significance for the field today from the point of view of developing a coherent theory of learning. This chapter gives an overview of the history of the field of mathematics education, particularly the influence of psychology, and a summary of the valuable contributions of theorists like Jean Piaget, Jerome Bruner and Zoltan Dienes Finally, the monograph includes a chapter on the pedagogical value of structural thinking by Jim Hirstein, who as a doctoral student at the University of Georgia intersected with Zoltan Dienes in the research meetings organized by Richard Lesh at Northwestern University in the mid 1970's (see the commentary of Lesh in Sriraman & Lesh, 2007). I hope this book will re-initiate an interest in the seminal work of Zoltan Dienes, particularly among those unfamiliar with his research and writings as well as contribute to future generations of mathematics educators appreciating his energy and monumental undertakings that led to the birth of our field of inquiry. This monograph is dedicated to him.

REFERENCES

D'Ambrosio, U. (2003). Stakes in mathematics education for the societies of today and tomorrow. In D. Coray, F. Furinghetti, H. Gispert, B. Hodgson and G. Schubring (Eds.): *One hundred years of l'Enseignement Mathématique*, (pp. 301-316). Monograph no.39. Geneva.

Dienes, Z. P. (1960). *Building up mathematics*. London: Hutchinson

Dienes, Z. P. (1973). *Mathematics through the senses, games, dance and art*. Windsor, UK: The National Foundation for Educational Research.

Dienes, Z. P., & Jeeves, M. A. (1965). *Thinking in structures*. London: Hutchinson.

English, L. D., & Halford, G. S. (1995). *Mathematics education: Models and processes.* Hillsdale, NJ: Lawrence Erlbaum.

Lesh, R., Post, T., & Behr, M. (1987). Dienes Revisited: Multiple Embodiments in Computer Environments. In I. Wirszup & R. Streit (Eds.) *Developments in School Mathematics Education Around the World.* Reston, VA: National Council of Teachers of Mathematics.

Lesh, R., Post, T. & Behr, M. (1989). Proportional Reasoning. In M. Behr & J. Hiebert (Eds.) *Number Concepts and Operations in the Middle Grades.* Reston, VA: National Council of Teachers of Mathematics.

Lesh. R., Behr, M., Cramer, K., Harel, G., Orton, R., & Post, T. (1992) *Five Versions of the Multiple Embodiment Principle.* ETS Technical Report.

Lesh, R., Cramer, K., Doerr, H., Post, T. & Zawojewski, J. (2003) Model Development Sequences Perspectives. In R. Lesh & H. Doerr, H. (Eds.) *Beyond Constructivism: A Models & Modeling Perspective on Mathematics Teaching, Learning, and Problem Solving.* Hillsdale, NJ: Lawrence Erlbaum Associates.

Skovsmose, O. (2005). *Travelling through education.* Rotterdam: Sense Publishers.

Sriraman, B., & Lesh, R. (2007). Leaders in Mathematical Thinking & Learning—A conversation with Zoltan P. Dienes. *Mathematical Thinking and Learning: An International Journal, 9*(1), 59-75.

CHAPTER 1

REFLECTIONS OF ZOLTAN P. DIENES ON MATHEMATICS EDUCATION

Bharath Sriraman[1] and Richard Lesh[2]

ABSTRACT

The name of Zoltan P. Dienes (1916–   ) stands with those of Jean Piaget, Jerome Bruner, Edward Begle, and Robert Davis as legendary figures whose work
left a lasting impression on the field of mathematics education. Dienes' name
is synonymous with the Multibase blocks which he invented for the teaching
of place value. Among numerous other things, he also is the inventor of Algebraic materials and logic blocks, which sowed the seeds of contemporary
uses of manipulative materials in instruction. Dienes' place is unique in the
field of mathematics education not only because of his theories on how mathematical structures can be effectively taught from the early grades onwards
using manipulatives, games, stories and dance (e.g., Dienes, 1973, 1987), but
also because of his tireless attempts for over 50 years to inform school practice
through his fieldwork in the UK, Italy, Australia, Brazil, Canada, Papau New
Guinea and the United States. Dienes' theories on the learning of mathemat-

Mathematics Education and the Legacy of Zoltan Paul Dienes, pages 1–19
Copyright © 2008 by Information Age Publishing

Figure 1.1 Zoltan P. Dienes and Bharath Sriraman, April 25, 2006, Wolfville, Nova Scotia.

ics have influenced many generations of mathematics education researchers, particularly those involved in the Rational Number Project and more recently those working in the models and modeling area of research. Dienes championed the use of collaborative group work and concrete materials, as well as goals such as democratic access to the process of mathematical thinking, long before the words "constructivism" and "equity" and "democratization" became fashionable. In this rare interview, Dienes reflects on his life, his work, the role of context, language and technology in mathematics teaching and learning today and on the nature of mathematics itself.

Sriraman: It is an honor to be able to talk with you. I really appreciate the invitation to visit.

Dienes: You have traveled so far...so I hope I am of some help. I can't have very much longer on this planet. So it's good you're here.

Sriraman: Your books have been very influential in my own work, many of your writings from the '60s, and especially the one you wrote when you were in Adelaide.

Dienes: With Jeeves, yes (see Dienes & Jeeves, 1965).

Sriraman: Yes, particularly the innovative experiments you set up which investigate reasoning about isomorphic structures such as groups....Do you still believe this is the way to teach mathematics, especially knowing that mathematics has become more and more applied in today's world compared to the '50s and '60s.

Dienes: Well! It depends on what you think in important and what one is after. Mathematics is characterized by structures, there is no denying this fact and in my opinion it is important to expose students to these structures as early as possible. This does not mean we tell them directly what these structures are but use mathematical games and other materials to help them discover and understand these structures. You have read about my theory of the six stages of learning (see Dienes, 2000b). And in this theory, the formalization stage comes at the very end.

Sriraman: As you know, there have been theorists who think that such topics are too difficult at earlier developmental stages—although your work indicates otherwise. Piaget, for instance thought this type of thinking (structural thinking) was only possible at the stage of formal operations.

Dienes: Children do not need to reach a certain developmental stage to experience the joy, or the thrill of thinking mathematically and experiencing the process of doing mathematics. We unfortunately do not give children the opportunities to engage in this type of thinking. One of the first things we should do in trying to teach a learner any mathematics is to think of different concrete situations with a common essence. (These situations) have just the properties of the mathematics chosen. Then... children will learn by acting on a situation. Introducing symbolic systems prematurely shocks the learner and impedes the learning of mathematics.

Sriraman: What are your thoughts on Piaget's theory?

Dienes: [Dienes gets up and retrieves a manuscript] I was working with Piaget's group of researchers at *Institut Rousseau* in Geneva I did not hear one consistent answer when I asked them what it means to be "operational"?...You can look at this manuscript and read what I asked Piaget.

Sriraman: (reading from Dienes' manuscript) "Is it so Monsieur Piaget, that a pre-operational child can operate on states to get to other states, but is unable to operate on an operator to get another operator, whereas an operational child can also operate on an operator, without having to think of the intervening states"

Dienes: Yes, and Piaget agreed with my definition (Laughing). You can read about my conception of operationality in children yourself. It is a bit different from Piaget's.

Sriraman: You mentioned pre-mature use of symbolic systems in the teaching of mathematics earlier. I agree that notation is used too early without children completely understanding what it is they are being forced to represent and symbolize. I know you spent a year at Harvard with Jerome Bruner. Do you care to talk about this?

Dienes: My emphasis was on the use of mathematical games with appropriate learning aids (manipulatives), work and communication in small groups with the teacher overseeing these groups....I did have arguments with Bruner and his followers on this subject. I even invented a term "Symbol Shock" [Laughing] and there was disagreement with my approach from his camp.

Sriraman: What got you interested in the teaching and learning of mathematics? You come from the background of being a mathematician...

Dienes: I explained it to some extent in that book (Dienes 2003). I thought it was strange how people didn't understand mathematics. What makes it so hard? Then I thought of things like...the distributive law for instance. It is very hard to explain this law to somebody who is not a mathematician, but you can invent some games which work in exactly the same way, which you can play. I thought why not try and see if you can do something like that with kids and see if they buy it. And they did.

Sriraman: From the point of view of a university teacher educator who wishes to make a structural approach to learning more common, how much mathematics do you think prospective teachers and teacher educators need to know before they can truly appreciate mathematical structures?

Dienes: The answer to your question depends on what you mean by how much mathematics? There are several things that are important. One needs to be able think logically. How much mathematics one studies...depends...

Sriraman: I think what I am trying to ask you is whether or not you think studying a lot of mathematics is important before starting to teach it.

Dienes: It really depends on the person. Some are able to grasp the fundamental ideas very quickly. So, if one doesn't study a whole lot of mathematics formally, but understands the material they have studied...it doesn't matter. The real problem occurs when one doesn't understand what mathematics is about in the first place and then tries to teach it. It is a question of depth.

You can learn mathematics simply as a utility and learn how to use it. That happened during the 18th and 19th centuries, during the Industrial revolution? It became necessary for people to read instructions, to do simple number work, because it was economically necessary. (But), all you had to do was learn certain tricks. To add, to multiply, get percentages, a little bit of fractions and so on. But, the situation today is different economically than it was say 150 years ago. It was good enough then to know just how to do the tricks. But it is not good enough anymore for doing the work we do now in most jobs. So, we need to know a little more mathematics. Now as to what type of mathematics we need to know, I suppose it doesn't matter very much because most mathematics you learn, if you understand it, will teach you a way of thinking... structural thinking. Thinking in structures, how structures fit into one another. How do they relate to each other and so on. Now, whether you learn that in Linear Algebra or in Infinite series or any other area.... As long as you get the idea of what mathematical thinking is like, you can apply it to all sorts of other situations.

Sriraman: Recently, there have been initiatives by Richard Lesh which are guided by your principles of learning. His research group uses *model eliciting activities* and *model development sequences* in much the same way that you used *concrete embodiments* and *multiple representations*. But, his work focuses on simulations of "real life" situations more than on concrete manipulatives. Students often work in small groups, just as in your work; and, their work continues to focus on structure. What do you think of this approach?

Dienes: Well, it is good to hear that others are making use of my learning principles. I emphasized small group work long before it became popular.

Sriraman: Do you think that real world contexts are relevant?

Dienes: Context is very important. In the work I've done in different parts of the world, I've always tried to put things in practical terms. It somewhat depends on the local culture in which you are operating. It wouldn't be the same in the United States as in India or China or New Guinea.... In New Guinea, I came across a tribe in whose language there was nothing for the concept of "either/or".... How do you teach logic if you don't know about "either/or". So I had to work out a way of making sure the kids understood. I did this by [tapping my arm] which meant it was correct; and, this [nodding head] meant no. So, with attribute blocks, a child would produce an answer to a question, and I would say [tapping my arm] that it was okay, and another child would produce another block, a different answer to the same question which was okay and I would again say [tapping my arm] that it was okay. That flabbergasted them. How could the answer be either this block or that block.... This was how I managed to teach the notion of "either or" in a culture where they had no words for such a concept.... In New Guinea, there is no "either/or" because the tribal system was so strict. You do THIS, and under these circumstances, you do THIS [under other circumstances]. And that's it. And God help you if you don't [Laughing].

Sriraman: Yeah, they had a uni-modal logic. That's interesting

Dienes: Yes [laughing]. So we can't really lay down the law to what should be in a teacher education program. It depends on the local situation. A set of tools that one learns can become completely useless in a different situation, and this can happen very fast..

Sriraman: I agree that the context will have to be built according to the reality in which students are situated. For instance, I have been reading the work of a researcher in the Chicago area, who has taught in a Chicago public school in which the children are pre-dominantly from the community of Mexican immigrants. This researcher is inspired by the work of Paulo Freire, the Brazilian...I don't know if you know about his pedagogy for social justice. The whole point this researcher makes is that the mathematics taught to these particular children needs to be socially relevant and promote a critical awareness of the reality in which they are living.

BD: [Bruce Dienes (BD), the youngest son of Zoltan Dienes enters]. I made copies of these articles that Zoltan published in the New Zealand Mathematics Magazine (see list of references for these particular articles).

Sriraman: Thank you, I have been trying to get some of these for a long time. Please join us.

Sriraman: In this pedagogy for social justice, mathematics is used to make sense of their reality. They use projects with real world data like mortgage approval rates in bigger cities according to race; the misinformation or distortion of land mass given in older maps using the Mercator projection. Interestingly these things came out during the peak on colonization. Then there are other projects like using the cost of a B-2 bomber to compute how many poorer students in that community could be put through university. It seems that in this approach the goal is to impact social consciousness and larger issues. Do you have any thoughts on critical thinking in the mathematics classrooms?

Dienes: I do understand what you are saying about socially relevant mathematics. As long as the problem engages the students, allows them room for play, and getting through the representational stage with the experience of multiple embodiments, it doesn't matter what types of mathematics we are dealing with. I assume these children are older.

Sriraman: Yes, this researcher in Chicago was teaching at a middle school.

Dienes: What I have been doing for over 50 years is not so much outside social issues but critical thinking about what mathematics is and what it can be used for and to have it presented as fun, as play, and in this sense it can be self motivating because it is in itself a fun activity. I have critiqued mathematics being presented as a boring repetitious activity as opposed to a way to think. So it is not so much critical thinking of social issues but as a way to train the mind, understand patterns and relationships, in ways that are playful and fun.

BD: Zoltan used to do a lot when making presentations is to intentionally make mistakes, encouraging students to correct him, to challenge authority. To not assume that whatever you are told is correct. His goal was to build a social environment where students don't have to lean on the teacher or on the tricks being taught. So his goal has been to address a more fundamental learning process as opposed to using issues to motivate it.

Sriraman: A lot of the reform based curricula are data driven, and emphasize modeling activities. The data is usually taken from reality and the mathematics is motivated from the need to make sense of this data. The mathematics usually is the study of various families of functions which can be used to model the data. One of the stories I remember from your book is when you had to teach partial differential equations (PDE's) to a group of engineering students in England during the war (WWII). And you were trying to create examples which would make the mathematics interesting to these students.

Dienes: I remember that, having to teach partial differential equations to engineering students. This couldn't have been of any possible use to these engineers whatsoever because what they would be interested in would be to determine what the arbitrary functions are which are part of a solution, not so much solving the equation. There are so few PDE's that can actually be solved [laughing] .One of the examples I concocted was being chased by a bull and having to determine using PDE's, conditions that would allow the person to escape [laughing]. In this case, I did not see the point of teaching methods of solutions to engineers with cooked up nice equations....which they would never have to deal with in their profession. Engineering situations are usually more messy and not as nice as those prescribed in the books.

Sriraman: Do you think your being exposed to so many languages as a child has something to do with your fondness for mathematics and the proclivity for thinking in structures?

Dienes: You are asking whether being brought up multilingual has a connection with doing mathematics. Well I am only one example so one cannot deduce from one example. There are some conflicting research findings on multilingualism and the development of intelligence for example. I mean, I never thought anything about the priority of one language as opposed to another one. You know, being brought up multilingual you think of this as a minor problem. I don't know if you remember the bit about when I was quite young, we would go to Czechoslovakia for our holidays and we would climb up our tree house and listen to the mathematics talks without them knowing we were there. They were all discussing problems of the day including mathematics and philosophy. That is partly what got me interested in understanding the mathematics. And when they were talking about philosophy, I thought aren't they clever, these guys, that just the process of thinking can be turned into fun [Laughing]. Just thinking about something can be fun. It's great...you don't really have to do anything. [Laughing]

Sriraman: I asked you the question about natural language because I noticed in the book, that at a very early age, you and your brother were able to create your own language. You came up with a list of nouns, a grammar and a syntax. Mathematics has some of these elements when one looks at if from a linguistic point of view.

Dienes: Have you looked at the Ruritania on my website?

Sriraman: Yes, I have. The adventures of Bruce and Alice. It is very clever, especially the creation of a new language.

Dienes: The vocabulary is made up. There is a certain logical relation between concepts which come out in the construction of the words. I don't explain that in the website. It is just there for somebody to discover. I have often thought that maybe this idea of constructing an artificial language could be a form of learning for the future. Like 50 years ago, the multibase blocks I brought in were regarded as absolute nonsense. Why did you do that? How could you possibly think of that as being of any use? Yet, now some people have finally understood. In fact those teachers that actually use it wouldn't want to go back. They've realized that it actually does teach children place value, the idea of the power as the exponent. I thought that something to do with expressing what you want to express in a language. You could think about what are the minimum conditions for a language to allow you to express what you want to express. You obviously have to have some kind of grammar, some kind of relationship between the words. What is the minimum core that you have to have. If you are able to construct a minimum core language like that, maybe it is something that will help you to understand different parts of the world, different cultures and different languages. Now, I just think this may happen in about a 100 years time, that someone will finally dig up this website [laughing] and say oh, 100 years ago there was this guy called Dienes, who was already thinking about it. Now we think it is obvious. But, a 100 years ago nobody took any notice of it [laughing]. It is in there, in that website. Somebody one day will realize the power in it.

Sriraman: The reason I asked you about the connection between language and mathematics is because I personally have seen language as being very structural. Growing up in India, I picked up 4 different languages very early. After that, it was quite easy to learn languages like Urdu and Farsi because structurally they were the same to others I already knew. The same goes with German and knowing English makes many other Indo-European languages in that family quite accessible. Once one discovers the structure, it is merely a matter of filling in the vocabulary and other irregularities in syntax.

Dienes: Yes, that was pretty much the case with me as well. I see the point you're making.

BD: Like you and Zoltan said, by being forced to learn multiple languages, one is forced to be conscious of the structures and notice it in the next one and learn it. If people have learned only one language, its like a fish in the water...you are completely unaware of the structure. Zoltan uses multiple embodiments as a learning strategy. When you do something with dance, you do it with blocks, and all of a sudden, as it is like for people like you and Zoltan with languages, all of a sudden multiple embodiments brings into consciousness the structures and patterns. I can see where you're coming from when you ask about the link to learning languages.

Dienes: The structural features one recognizes from these multiple embodiments—this brings out the essence of abstraction. Symbolism can be thrown in at this advanced stage, not earlier. The reason mathematics is boring in schools is because no real mathematics is taught in schools [Laughing]

BD: When I first went to McGill, I did one year of pure mathematics and then decided to go off in a different direction. In that first year, it was like, that's what it is all about. It was as if there was a secret society which only got to know these things, these underlying structures. You had to be indoctrinated into honors courses in mathematics before they told you what mathematics was really all about. For 12 years until that point, I had really learned nothing. I didn't truly know what multiplication was until I took an honors course in Abstract Algebra. So, part of the reason, students get frustrated is because they aren't really being taught anything. The radical thing that Zoltan tried to do with young children is to try and teach them some mathematics, the process of thinking mathematically....It worked, and everybody got up in arms...You can't do that! Because that would mean that the teachers and everyone else would really have to know it.

Dienes: [Laughing]...and the secret society would no longer be secret.

Sriraman: There is plenty of research now which indicates that mathematics is the key to opening up numerous professions for the children in schools, new opportunities. In order to get into the sciences or any applied fields, mathematics has historically served as the sieve or the gatekeeper to peter our people who supposedly don't fit into this secret society. It still is.

Dienes: You're absolutely right. I have seen this happening in different parts of the world.

Sriraman: If one were a cynic, I suppose you could think of mathematics as a means of keeping society stratified, and promoting the status quo and inequity in place.

Dienes: There are many such mechanisms, mathematics is one of them. You know reading used to be regarded as the preserve of the scribes. You didn't want the masses knowing how to read, that would be bad. It would disturb society.

Sriraman: It is clear from your memoir that you took to incorporating technology about as quickly as it was being invented. These days it seems that the younger generation is very adept at adapting to new technologies and becoming experts at using them. It is not uncommon to go to a public library in North America and see a 3 or 4 year old comfortably using the mouse to navigate their way through games etc. How do you think technology can be used to our advantage in the teaching and learning of mathematics?

Dienes: Like everything else, every major change will have an effect on just about everything else you do in life. I mean, just like running the house has changed because of technology, driving a car, everything has changed. So, it obviously has to have some kind of effect on the teaching of mathematics....I don't know. I mean you are in the next generation, and I am already going out. I've done my bit so to speak and what I've done is probably not good enough. You need to develop it further.

Sriraman: The reason I asked you this was because there is always some form of controversy or another on the use of technology in a mathematics classroom. The purists versus the reformists, things like technology hindering the learning of mathematics.

Dienes: [Laughing] Well, you see it puts them out of business. If the calculator or other computer technology can now perform the calculations, they can no longer teach the tricks and hide behind it. The focus will now have to be on understanding the tricks.

BD: Creating structured materials was what Zoltan was doing with the logic blocks etc, The nature of the materials determined the nature of play to some degree. So even if we gave them no instructions at all, they could learn some principles as a result of interacting with the materials. The computer is ideal for setting up structured environments. That is what programming is. What you need is a structured environment with principles built into it, but open enough that you can interact and change with it. This is why role playing games in computing environments are so popular. The question is to set up the right balance between interactivity and structure to whatever stage is appropriate for a learner. The biggest problem is that they are generally one on one because they've been designed for office use. It doesn't mean that is how they have to be used. I mean, why should a screen be this size by this size and placed the way it is? Why can't we have a table that is a screen so people can sit around and manipulate the environment together, interact with touch. We need to radicalize the interface between student and computer, not take an office machine and put it in the classroom and hope that it will work. Take the principles of computing technology, redesign things from scratch to make them suitable to teach.

Dienes: You'll be putting a lot of people out of business doing that [laughing].

Sriraman: What are your views on the nature of mathematics? Do you think mathematics is a human creation or do you think it is discovered?

Dienes: Well, I think this is a non-problem because we will never solve it. But we can talk about it [laughing]. Yes. Well, I have an idea that certain things are so whether we think them or not, so in that sense I am with Plato. I mean $2 + 2$ is always going to be 4 in any system you are likely to design. There are certain core elements that must be so, whether we think so or not. When you discover, in quotes, mathematics. Are you creating the mathematics or are you discovering what is already there? I'd rather fancy you're discovering. Mathematics being the only branch of knowledge for which such a sentence can be said! If you take something like Physics, there are already certain assumptions you make about the nature of matter, which may be open to question before you can use mathematics to deal with a problem in physics. Like quantum computing, you have to first know something about quantum theory, assumptions about matter etc to even know what people are talking about. Do you read Scientific American? Yes...So you've read the recent arguments on this issue. One of my uncles used to be a research biochemist at Harvard. He gave me Scientific American for a year, 30 years ago or more, and at the end of the year I didn't think I could do without it and I've had it ever since [Laughing]. So we leave the non problem. I really don't think it is a problem [laughing]. Playing with new forms of mathematics enables us to reframe how we look at the universe, and find things we may not have found had we not been able to reframe it that way. It is a bit of both.

BD: I spent some time visiting the classes in which Zoltan did his teaching. What I noticed was not so much the mathematics.

You know his second degree was on the psychology of learning mathematics. So it was more about the learning than the mathematics. In fact in some of his classes in Quebec he would have a mixture of mathematics, language and art in the same classroom, with different learning stations where students could choose what they wanted to work on. Often these classes had multiple grades, so the older ones were teaching the younger ones. The other thing he did was he never ever set up competitive games. The games were always things that didn't work unless you worked together. His strategy was always to focus on the nature of learning. How do we create an environment in which people learn to cooperate, to have fun, and have choices and power over their own learning experience? And create an atmosphere where learning is empowering. It is pretty much the kinds of things you were talking about.

Sriraman: It seems to me, based on reading your works from the '60s, that there was certainly a Piagetian influence. Perhaps you remember the book Piaget wrote with Beth, where he claims there is a correspondence between the evolving cognitive structures within an individual's mind to those of the structures espoused by the Bourbaki, namely structures of order, algebraic structures and then topological structures. Piaget was never able to prove his claim. Do you have any thoughts on this?

Dienes: I know what you are talking about but I have never been very clear on that. The one time that Piaget came and listened to my teaching kids in Geneva. Well! Piaget himself and some of his students including Inhelder I think. All of them came and I showed them how to teach complex numbers with a story. Not with manipulatives mind you but by simply manipulating a story line and I remember that his students were flabbergasted and when they came out I remember them saying. *C'est ne pas possible! C'est ne pas possible!* You know as if this didn't really happen. And I asked Piaget. *Et vous Monsier Piaget, Qu'est-ce que vous en pensez?* His reply was *Tiens, c'est très interéssant* [Laughing]. So that is the extent to which he commented on it. But the fact that you could have a story and get the learning going was not in his theory you see. So you see it was something neither he nor his students could handle. Piaget was a god you see and he can't be wrong. Yet, here was something his theory couldn't explain.

Sriraman: There is new evidence now that many of the stages espoused by Piaget in this theory are not set in stone. Children are capable of doing things that could only occur in later stages according to his theory, especially if they are given child appropriate materials in contrast to the types of materials he used.

Dienes: I have always been more practical in my theorizing than people like Piaget or Bruner. Let's stick to the facts and see what is possible. A lot of people that have seen me in the classroom with children have said that the things that I showed them were possible were things they never thought could be. When they saw actual learning taking place at my instigation with materials, stories etc.

Sriraman: Do you think it is possible to have some sort of a global theory of learning?

Dienes: I do not know.

Sriraman: Do you know why your theories of learning did not catch on as much as they should have?

Dienes: The answer to that is simple. They were politically unpalatable [Laughing].

BD: I can give you an example of what he is saying based on his experiences in Brazil. You know they would always take him to the show schools, and he said, why don't we go down to the Barrio. They said, No, you don't want to go there. The children there are not so intelligent. In any case he went to these schools and did some work down there and of course they learned just as well and as quickly and he was never allowed to go back. This was politically incorrect information with the responsibility to provide these children the same kind of training and resources the privileged schools were getting.

Dienes: I have never believed in a curriculum for young children. What matters is that children learn how to think. So trying to convince principals and people and others in charge that this type of learning adapts to the needs of the students doesn't match their conception of how learning occurs. I have not been one to patronize the establishment you know.

I remember being taken into one of these really poor schools in Brazil. Yes, you went into the classroom and half the classroom was occupied with broken down tables and chairs and what not piled up in a pile. And each child has a piece of paper about this size [indicating the palm of his hand] and the one pencil and that was their material. I was taken in there and asked what can you do with these? I said this is a challenge [laughing]. I said, first thing we'll do is see what's in that corner. Let's take those broken chairs down and see what we can do with them. Oh! Shall we? Yes!. I got the kids to sort out the mess in that corner. Any tables we could make into tables we made, with screws and nails. And we also made some actual mathematics materials with the stuff and started learning mathematics. All we needed was that thrown away stuff. Previously all they had thought about mathematics was this one little piece of paper and one pencil. [Laughing].

Tessa Dienes: The sad end of this story is that they brought us a box of expensive chocolates at the end and gave us as a present. And the first bit of information was that we mustn't share them with the children. Oh no, we shouldn't do that. Goodness knows where they got the money to buy the chocolate. We weren't able to share them.

Sriraman: What are your thoughts on constructivism, which became popular when people in mathematics education started to re-discover Piaget and Vygotsky's writings in the '80s?

Dienes: My answer to that is simple really. These things were practiced in my classrooms long before people invented a word for it and I am sure there are people out there doing similar things for children. I do not care very much for isms, be it constructivism or behaviorism or any other ism. What really matters is that actual learning can take place with the proper use of materials, games, stories and such and that should be our focus. Ultimately, have they learned anything that is useful and made them think?

Sriraman: You seem to be getting tired now. I would be happy to stop here. I thank you again for your time.

Dienes: I hope I have been of some use. I'm not very young anymore as you can see. Let us eat some lunch.

COMMENTARY BY RICHARD LESH

I first met Zoltan Dienes in 1974, during the first in a series of annual research mini conferences that I used to hold at Northwestern University. These mini-conferences later led directly to the founding of both PME/NA and the NCTM Research Presessions. At that meeting, one evening, I had the wonderful opportunity to sit around talking long into the night with Dienes and Bob Davis. It was an opportunity I'll never forget.

Like Davis, Dienes was educated as a mathematician. But, both Dienes and Davis were Renaissance men of the highest order—and visionaries whose expertise spanned a wide range of fields both practical and theoretical. Furthermore, both considered mathematics education to be a place for scientific inquiry—rather than a venue for spouting personal prejudices. So, both conducted extensive studies about the psychology of mathematics learning and problem solving; and, both hobnobbed regularly with people such as Jean Piaget, Jerome Bruner, and other giants whose thinking shaped modern theories of cognitive science.

Dienes, in particular, introduced many ideas that only now, thirty years later, are beginning to be appreciated for their power and beauty. Examples follow.

Like Piaget, Dienes emphasized the fact that mathematics is the study of structure—and that many of its most important concepts and processes have meanings which depend on thinking that is based on conceptual systems-as-a-whole. For instance, using a wide range of creative tasks, Piaget demonstrated the inherent systemic nature of (a) unit concepts whose meanings depend on invariance properties (with respect to a system), (b) relation concepts whose meanings depend on properties such as transitivity (with respect to a system), and (c) other properties such as those which depend on patterns or regularities (of a systems)—or on the maximization, minimization or stabilization of properties with a system. He demonstrated that statements of belief often are emergent properties of systems of belief, that statements or principle often depend on systems of principles, and that these systems need to function as systems-as-a-whole before the relevant concepts, principles, and beliefs attain their intended meanings. Finally, he demonstrated what children's thinking is like before relevant conceptual systems-as-a-whole have begun to function as systemic wholes; and, he demonstrated some of the most important processes which influence development. But, one of the things the Piaget did *not* do, and that Dienes definitively *has* done, is to recognize that mathematics is not just about structure; but, even more importantly, it is about isomorphism, homomorphism, and more generally structural mappings among structures. Furthermore, relevant conceptual systems are molded and shaped by the external systems they are used to interpret, and that beyond entry-level mathematical systems usually need to be expressed using some external media—or embodiment—if they are to function properly. Therefore, Dienes not only studied a phenomena that later cognitive scientists have come to call embodied knowledge and situated cognition—but he also emphasize the *multiple embodiment principle* whereby students need to make predictions from one structured situation to another. And, he also emphasized the fact that, when conceptual systems are partly off-loaded from the mind using a variety of interacting representational systems (including not only spoken language written symbols, and diagrams, but also manipulatives and stories based on experience based metaphors), each representational) that every such model is at best a useful oversimplification of both the underlying conceptual systems being expressed and the external systems that is being described or explained.

Thus, Dienes' notion of *embodied knowledge* presaged other cognitive scientists who eventually came to recognize the importance of *embodied knowledge* and *situated cognition*—where knowledge and abilities are organized around experience as much as they are organize around abstractions (as Piaget, for example, would have led us to believe), and where knowledge is distributed across a variety across tools and *communities of practice.*

Dienes was an early pioneer in what is now coming to be called *sociocultural perspectives* on learning and problem solving. This can be seen in the fact that the learning activities that he designed nearly always involved groups of students; and, they were not just activities that were written for isolated individuals which happened to be approached by a group. Instead, Dienes' activities explicitly build in characteristics which demanded that the "learner" needed to be a group. Thus, for Dienes, and unlike what many modern sociocultural theorists suggest, the learning community was not simply an entity which allowed individuals to learn from others—or to learn by adopting cultural artifacts from the community as a whole. Instead, for Dienes, the learner often *is* a group; and, this fact is becoming increasingly important to notice in the knowledge economies of an information age— when learning organizations need to continually adapt to rapidly changing circumstances. Finally, Dienes brand of situated cognition was not simply one in which knowledge development is considered to be task specific. Instead, Dienes emphasized the fact that, when conceptual tools are developed, they usually are designed to be sharable (with others) and reuseable (beyond the initial situation in which they were needed).

As Sriraman mentions in this interview, Dienes' theories strongly influenced a number of the most productive research programs in mathematics education. One prime example is the series of projects that came to be known collectively as *The Rational Number Project* (Lesh et al., 1992; Post et al., 1992). Another, is the work of Kaput and others who focused on technology-based learning environments and on multiple-linked representational media (Lesh et al., 1987). And, more recently, extensions of Dienes' ideas play central roles in current research on *Models & Modeling Perspectives* on mathematical thinking and learning (Lesh et al., 2003) and on teacher development (Doerr & Lesh, 2003).

NOTES

1. The following interview took place at Zoltan Dienes residence in Wolfville, Nova Scotia on April 25, 2006. The authors appreciate Bruce Dienes' assis-

tance in some portions of the interview as well as the hospitality of Sarah and Tessa Dienes.

2. Richard Lesh kindly agreed to construct a discussion/commentary at the end of this conversation

REFERENCES

Dienes, Z. P. (1960). *Building up mathematics*. London: Hutchinson.

Dienes, Z. P. (1963). *An experimental study of mathematics-learning*. London: Hutchinson.

Dienes, Z. P. (1964). *The power of mathematics*. London: Hutchinson Educational.

Dienes, Z. P. (1965). *Modern mathematics for young children*. Harlow, UK: ESA Press.

Dienes, Z. P. (1971). An Example of the Passage from the Concrete to the Manipulation of Formal Systems. *Educational Studies in Mathematics* 3. 337–352.

Dienes, Z. P. (1973). *Mathematics through the senses, games, dance and art*. Windsor, UK: The National Foundation for Educational Research.

Dienes, Z. P. (1987). Lessons involving Music, Language and Mathematics. *Journal of Mathematical Behavior*, vol. 6, 171–181.

Dienes, Z. P. (1999). Getting to know the rotations of the cube. *The New Zealand Mathematics Magazine*, vol. 36, no. 3, pp. 14–27.

Dienes, Z. P.(2000a). Logic Axioms. *The New Zealand Mathematics Magazine*, vol. 37 no. 1. pp. 21–35.

Dienes, Z. P. (2000b). The Theory of the Six Stages of Learning with Integers. *Mathematics in School*, vol. 29, no. 2, pp. 27–33.

Dienes, Z. P. (2001a). Circular villages. Part 1. *The New Zealand Mathematics Magazine*, vol. 38, no. 1, pp. 23–28.

Dienes, Z. P. (2001b). Circular villages. Part 2. *The New Zealand Mathematics Magazine*, vol. 38, no. 2., pp. 25–33.

Dienes, Z. P. (2003). *Memoirs of a maverick mathematician* (2nd ed.). Leicester, UK: Upfront Publishing.

Dienes, Z. P. (2004a). Six stages with rational numbers. *The New Zealand Mathematics Magazine*, vol. 41. no. 2, pp. 44–53.

Dienes, Z. P. (2004b). Points, Lines and Spaces. *The New Zealand Mathematics Magazine*, vol. 41, no. 3, pp. 33–42.

Dienes, Z. P., & Golding, E. W. (1966). *Exploration of space and practical measurement*. New York: Herder and Herder.

Dienes, Z. P., & Jeeves, M. A. (1965). *Thinking in structures*. London: Hutchinson.

Doerr H. & Lesh, R. (2003) *A Modeling Perspective on Teacher Development*. In R. Lesh & H. Doerr (Eds.) *Beyond Constructivism: A Models & Modeling Perspective on Mathematics Teaching, Learning, and Problems Solving*. Hillsdale, NJ: Lawrence Erlbaum Associates.

Lesh, R., Post, T., & Behr, M. (1987). Dienes Revisited: Multiple Embodiments in Computer Environments. In I. Wirszup & R. Streit (Eds.) *Developments in School Mathematics Education Around the World*. Reston, VA: National Council of Teachers of Mathematics.

Lesh, R., Post, T. & Behr, M. (1989). Proportional Reasoning. In M. Behr & J. Hiebert (Eds.) *Number Concepts and Operations in the Middle Grades*. Reston, VA: National Council of Teachers of Mathematics.

Lesh. R., Behr, M., Cramer, K., Harel, G., Orton, R., & Post, T. (1992) *Five Versions of the Multiple Embodiment Principle*. ETS Technical Report.

Lesh, R., Cramer, K., Doerr, H., Post, T. & Zawojewski, J. (2003) Model Development Sequences Perspectives. In R. Lesh & H. Doerr, H. (Eds.) *Beyond Constructivism: A Models & Modeling Perspective on Mathematics Teaching, Learning, and Problem Solving*. Hillsdale, NJ: Lawrence Erlbaum Associates.

Post, T., Behr, M., Lesh, R., & Harel, G. (1992) Learning and Teaching Ratio and Proportion: Research Implications. In D. Owens (Ed.) *Mathematics Education Research in the Middle*. Macmillan: New York, NY.

CHAPTER 2

SOME PROBLEMS WITH LOGIC BLOCKS

Suggested by Zoltan Dienes

TWO VALUES FOR EACH VARIABLE

Take out all the blocks with two shapes and with two colours,
For example you could use all the
squares and circles, red as well as yellow.

Let us call the properties red, square, big and thick the strong properties
and the properties yellow, circle, small and thin, the weak properties. The
"weakest" block will be the small thin yellow circle. We shall call this the
neutral block. We shall have 15 blocks, apart from the neutral one.

1. How to "add" two blocks?

    Choose any two of your fifteen blocks. Call them A and B

    Place the neutral block next to one of these. Say next to block A.

    Notice precisely what the differences are between A and the neutral
    block.

Mathematics Education and the Legacy of Zoltan Paul Dienes, pages 21–35
Copyright © 2008 by Information Age Publishing
All rights of reproduction in any form reserved.

Now find the block which has these exact differences between it and the block B.

The block you have found, say the block C, will be called the sum of A and B.

2. Make a circle of the 15 non-neutral blocks.

Put the fifteen blocks out in a circle in any order.

Now rearrange them round your circle so that the red ones and the yellow ones follow this sequence:

red, red, red, red, yellow, yellow, yellow, red, yellow, yellow,
red, red, yellow, red, yellow

Then exchanging red ones with red ones and yellow ones with yellow ones, make sure that the shapes come in a similar sequence to the one above, namely as

square, square, square, square, circle, circle, circle, square,
circle, circle, square, square, circle, square, circle

Having done the above, also arrange the big ones and the small ones, as well as the thick ones and the thin ones in such a way that all four properties follow the cycle:

strong, strong, strong, strong, weak, weak, weak, strong, weak, weak,
strong, strong, weak, strong weak

To make the big-small cycle, you must only exchange blocks of the same colour and shape. Finally to make the thick-thin cycle, exchange only blocks of the same colour, shape and size.

Here is another way of making the circle:

Start the circle with the four blocks each having just one strong property. These will be your blocks 1, 2, 3 and 4. To get block 5 add blocks 1 and 2. To get block 6 add blocks 2 and 3. To get block 7 add blocks 3 and 4. Go on like this until you have placed all the 15 non-neutral blocks.

Or you can start with any four blocks such that not one of them is the sum of any two of the other three, or the sum of all the other three. The "one strong property" for each of the four starting blocks is the easiest way of making sure of the above.

Now check how the strong and weak properties alternate as you go round the cycle.

Do you get the same sequence as before? Do you get it with all four properties?

There is a solution of the problem on the diagram provided.

3. Regularities hidden in the circle.

    You can find some interesting regularities in this circle of blocks. For example:

    a. every thin one is followed by a small one, every thick one by a big one
    b. every small one is followed by a circle, every big one by a square
    c. every circle is followed by a yellow one, every square by a red one
    d. every yellow square or red circle is followed by a thick one, and every yellow circle or red square is followed by a thin one.

    If you look at two consecutive blocks, there are some rules for finding the next block. For example:

    a. If the two blocks have different colours, the third block is big, if they have the same colour, the third block is small.
    b. If the two blocks have different shapes, the third block is thick, if they have the same shape, the third block is thin.
    c. If the two blocks have the same colour and the same thickness, the third block is a circle. If the two blocks have the same colour but different thickness, the third block is a square. If the two blocks have different colours and different thickness, the third block is a circle. If the two blocks have the same colour but have different thickness, the third block is a square.
    d. Give each block a point for each strong property it has. Add the points for size, thickness and colour in your two consecutive blocks. If this number is odd, the third block is red, if it is even, the third block is yellow.

    Maybe you can find a simpler way of expressing the last rule!

    Now try and find a rule for getting not the next block to any block, but the next block but one. Then find a rule for the previous block to any block. Where will you find the sum of any two consecutive blocks?

    What would happen if there were three colours, three shapes, three sizes and three thicknesses? You would have to invent an "addition" for these blocks. Perhaps you could start with just one thickness, but three shapes, three colours and three sizes? Or just one size to begin with?

Figure 2.1 is the picture of the circle of blocks that I made.

Figure 2.1

It "begins" on the right with the small thin red circle.

We can decide to list the properties, for example, in this order :

size, thickness, colour, shape

and write 1 for a strong property and 0 for a weak property, then the cycle can be described like this:

0010, 0100, 1000, 0001, 0110, 1100, 1001, 0111, 1010, 0101,
1110, 1101, 1111, 1011, 0011

3. More circles

You could make your own circle, find the rules which give you the next block after any given block and then give these rules to a friend who would have to make your circle without looking at the circle you have made. For example there is one circle with these rules:

a.   big yellow or small red → big next
            small yellow or big red → small next
b.   red → thick next
                   yellow → thin next
c.   square → red next
                   circle → yellow next
d.   odd number of strong properties → square next
     even number of strong properties → circle next

Now construct the circle. Figure 2.2 shows the solution.

Figure 2.2

When I wrote the instructions for making the cycle of fifteen blocks by arranging all the strong-weak sequences to be the same for each property, I had never actually done it myself. So I thought I had better have a go. Here is the arrangement that I got:

0011, 0010, 1001, 0101, 0001, 1011, 1100, 0100,1010, 0111,
1000, 1110, 1101, 1111, 0110

If you look at any run of five of the above, you will find that adding the first two blocks you always obtain the fifth. Is this just fluke? Or will it always happen if we follow the instructions for making the circle of blocks? And if so, can you prove it?

You can think of "going round the cycle" as a multiplication. Let us use our first cycle. We can appoint one of the blocks as a "neutral multiplier" For example we could appoint 0010. This would mean that "multiplying" by 0010 would leave us with the same block. For example

$$1010 \times 0010 = 1010$$

Multiplying by 0100 would lead to the next block in the circle. Multiplying by 1000 would lead us to the next block but one. The product of a block A by another block B is as many spaces along from B as A is from the neutral. You can easily verify that for any blocks A, B, and N, it will always be true that

$$A \times N + B \times N = (A + B) \times N$$

This means that the number of spaces between A, B and A + B will be the same as the number of spaces between A × N, B × N and (A + B) × N.

So with our blocks we have constructed a "mini-algebra."

Mathematicians call such a mini-algebra a field.

So why don't you look up "fields" in an algebra book?

See if you can make up some mini-algebras or "fields" of your own

Here is another circle. In this circle

odd number of strong points means a big one follows,
even number of strong points means a small one follows.

Odd number of non-colour points means a red one follows
Even number of non-colour points means a yellow one follows.

A big one is followed by a thick one
A small one is followed by a thin one.

A red circle or a yellow square will be followed by a square
A red square or a yellow circle will be followed by a circle.

You can check that for each property the strong and the weak follow the pattern

strong, strong, strong, strong, weak, weak, weak, strong,
weak, weak, strong, strong, weak, strong, weak

Figure 2.3 is a picture of this circle:

Figure 2.3

4. Find the triads.

   You might have noticed that if for three blocks A, B and C: A = B + C, then also B = A + C and C = A + B. In other words, each block is the sum of the other two. Such a set of three blocks we shall call a triad. You will already know a certain number of triads.

   For example in any run of five consecutive blocks the first two, together with the fifth one will always be a triad. You can obviously select a run of five in fifteen different ways. This gives you fifteen triads.

   You can also go round the circle selecting every fifth block, which will also yield a triad.

   This can be done in five different ways. So far we have twenty triads.

   How many triads are there?

   Try to find all of them.

EXTENSION TO MORE VALUES FOR A VARIABLE

From a box of logic blocks take out all the rectangles (but *not* the squares!) as well as all the big thin blocks. You will be left with 27 blocks. You will have 9 circles, 9 squares and 9 triangles. Among these there will be 9 yellow ones, 9 blue ones and 9 red ones. Also 9 of them will be small and thin, another 9 will be small and thick and another 9 will be big and thick. I shall refer to these as the values of the dimension variable.

In what follows I shall call the small thin ones just thin, the small thick ones just thick and the big thick ones just big.

   I shall refer to the properties red, triangle and big as strong properties.

   The properties blue, square and thick will be our middle properties.

   Finally thin, yellow and circle will be our weak properties.

   The thin yellow circle will be the weakest block and so I shall adopt it as the neutral block in what follows.

1. How to add two blocks.

   We shall "add" blocks by "adding" each of the properties separately. The dimension of one block is "added" to the dimension of the other block. The colour of one block is "added" to the colour of the other block. The shape of one block is "added" to the shape of the other block.

These will be our "adding rules":

> any property + a weak property = the same property.

In other words, adding weak does not alter the "strength" of a property.

> middle + middle = strong, strong + strong = middle

> middle + strong = strong + middle = weak

Besides "adding" I shall use the operation that I shall call "swapping" with these rules:

> swap of middle = strong, swap of strong = middle,
> swap of weak = weak

2. Constructing a circle with 26 non-neutral blocks.

Let us start with the three blocks:

> thin yellow square, thin blue circle, thick yellow circle

and get the next block by adding the swapped first block, the swapped second block and the third block. Then we can continue, finding the fourth block after any three by the rule

> Swapped A + Swapped B + C = the next block after A, B, and C

We shall get this sequence with these rules of succession:

thin yellow square	thin yellow, thick red, big blue → thin
thin blue circle	thin blue, thick yellow, big red → thick
thick yellow circle	thin red, thick blue, big yellow → big
thick red triangle	
thin blue triangle	thin circle, thick square, big triangle → yellow
thick red circle	thin square, thick triangle, big circle → blue
thin red triangle	thin triangle, thick circle, big square → red
big red circle	
thick blue square	thin → circle
big yellow triangle	thick → triangle
big yellow square	big → square
big red square	
thick red square	OR
thin yellow triangle	

thin red circle	dim strength + colour strength
big yellow circle	gives dim strength of next block
big blue square	
thin red square	swapped dim-strength + shape strength
big blue circle	gives colour strength of next block
thin blue square	
thick blue circle	swapped dim strength
big red triangle	gives shape of next block
thick yellow square	
thick yellow triangle	
thick blue triangle	
big blue triangle	

Now you try and find the rules for the next block but one, as well as for a block that is a given number of blocks ahead of any chosen block.

You might have noticed that in any run of six consecutive blocks, the sum of the first block and the fourth block is always the sixth block or that in any run of nine consecutive blocks the sum of the eighth and the ninth blocks is always the first block. Try to find more "addings" like that.

It is worth noting that any two blocks which are thirteen blocks away from each other are each other's swapped versions. So once you have made the first thirteen blocks of your circle, you can just put down the other thirteen by swapping the ones you have already got, from the first one to the thirteenth one, of course in the same order.

The dimensions, the colours and the shapes each alter following the same sequence of strengths, but of course each cycle starts at a different point in the circle. Using W, M and S for weak, middle and strong respectively, the cycle here seems to be

S M S S S M W W S S W S W M S M M M S W W M M W M W

We can place three such cycles next to each other, "starting" each cycle at the appropriate place and we shall obtain our circle of blocks.

There are also rules for getting the third block after any two consecutive blocks.

These seem to be the following:

For the next shape we define the "score" as follows, the numbers in the brackets referring to block (1) and block (2) that we are considering. All numbers should be interpreted as mod 3 numbers. Remember weak = 0, middle = 1, strong = 2. (You might have worked out by now that what I called "swapping" is just a multiplication by 2 in mod 3) score = dim (1) + col (1) + col (2) + 2 × shape (1) + 2 × (shape (2)

score = 0 you get a circle for the third block

score = 1 you get a triangle for the third block

score = 2 you get a square for the third block

one block is thin, the other is thick or both are big
   you get a red one next

one block is big the other is thick or both are thin
   you get a yellow one next

one block is big the other is thin or both are thick
   you get a blue one next.

To get dimension of the third block the score is the sum of all the points, counting both blocks.

score = 0   you get a thin one next

score = 1   you get a big one next

score = 2   you get thick one next

Figure 2.4

The first column gives the strengths of the *dimension* variable, the second one gives the *colour* and the third one gives the *shape*. You will find the layout of the blocks on the next page (Figure 2.5).

Dimension	Colour	Shape
S	S	S start
M	W	M
M	W	S
M	M	S
S	M	S
W	W	M
W	M	W
M	W	W
M	S start	S
W	M	S
M	S	W
W	S	S
S start	S	W
M	M	M
S	W	S
S	W	M
S	S	M
M	S	M
W	W	S
W	S	W
S	W	W
S	M	M
W	S	M
S	M	W
W	M	M
M	M	W end
S	S	S
M	W	M
M	W	S
M	M	S
S	M	S
W	W	M
W	M	W
M	W end	W
M	S	S
W	M	S
M	S	W
W end	S	S
S	S	W
M	M	M

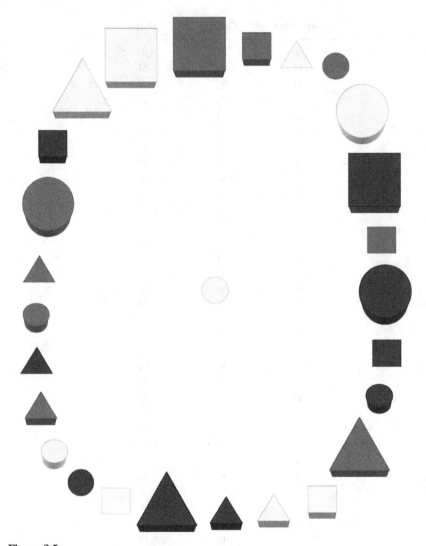

Figure 2.5

3. Back to the "shallow end"

If we want to simplify things, we can either the reduce the number of variables or the number of vales each variable can assume, or we can do both of these things. So we could examine what would happen if we made up a situation with two variables, each having three values and maybe another with three variable each having two values. These could be

The Montana Mathematics Enthusiast

2007 Monograph 1

International Perspectives on Social Justice in Mathematics Education

The Montana Mathematics Enthusiast

2007 Monograph 1

*International Perspectives on Social Justice
in Mathematics Education*

Edited by

Bharath Sriraman
The University of Montana

INFORMATION AGE PUBLISHING, INC.
Charlotte, NC • www.infoagepub.com

Library of Congress Cataloging-in-Publication Data

International perspectives on social justice in mathematics education /
edited by Bharath Sriraman.
    p. cm. – (The Montana mathematics enthusiast : monograph series in
mathematics education)
  Includes bibliographical references.
  ISBN-13: 978-1-59311-880-8 (pbk.)
  ISBN-13: 978-1-59311-881-5 (hardcover)
  1. Mathematics–Study and teaching–Social aspects. 2. Right to
education. 3. Educational equalization. I. Sriraman, Bharath.
  QA11.2.I67 2008
  510.71–dc22   2007047727

Permission to photocopy, microform, and distribute print or electronic copies may be
obtained from:
  Bharath Sriraman, Ph.D.
  Editor, *The Montana Mathematics Enthusiast*
  The University of Montana
  Missoula, MT 59812
  Email: sriramanb@mso.umt.edu
  (406) 243-6714

Printed in the United States of America

CONTENTS

FOREWORD

Social justice allows us not only to know what has been decided about ourselves and society (which is the objective of "re-productory" and imitative education), but calls us to participate in decisions about ourselves and society (which is the objective of creative critical education). This is what Paulo Freire had in mind, and is not different from my vision about the objectives of education, which is: to promote creativity, help people to fulfill their potentials, rise to the highest levels of their capabilities *and* to promote citizenship, transmitting values and showing rights and responsibilities in society. *But* being careful to promote neither irresponsible creativity—we do not want our students to become bright scientists creating new weaponry—nor docile citizenship—we do not want our students to accept rules and codes which violate human dignity. This book compiled and edited by Bharath Sriraman is an astonishing collection of scholarly articles from all over the world offering a kaleidoscope of perspectives of tremendous importance to mathematics educators interested in our shared concern to create a saner, equitable, and more peaceful society.

Ubiratan D'Ambrosio
São Paulo, Brazil

International Perspectives on Social Justice in Mathematics Education, page vii
Copyright © 2008 by Information Age Publishing
All rights of reproduction in any form reserved.

CHAPTER 1

ON THE ORIGINS OF SOCIAL JUSTICE

Darwin, Freire, Marx and Vivekananda

Bharath Sriraman
The University of Montana, USA

ABSTRACT

This article examines the fundamental reasons for educational research and practice in social justice from evolutionary, ideological and philosophical viewpoints. The tension between nihilistic[1] and empathetic tendencies within humanity's evolution is used to reflexively examine the origins and causes of inequity. The relevance of the works of Paolo Freire, Karl Marx, and Vivekananda for contemporary social justice research is examined.

WHY SOCIAL JUSTICE IN MATHEMATICS EDUCATION?

This ambitious book has finally reached completion and brings to fruition the hard work and initiatives of many individuals scattered across the globe. Editing and compiling this book has not simply been a learning experi-

International Perspectives on Social Justice in Mathematics Education, pages 1–9
Copyright © 2008 by Information Age Publishing

1

ence but one of increased awareness on the inequities and social injustices inherent within institutional and societal mechanisms and the complexities of addressing these issues within an educational context. Although the title clearly indicates this book is about international perspectives on social justice in mathematics education, in my view it is really a book about our attempt to create *Meaning*.

A nihilist would question: Why social *justice*? In other words, what is it about society and education today that is broken and needs fixing or needs to be ad*just*ed? It is a basic fact that life around us constantly reveals inequities such as rich versus poor; the educated versus uneducated; those in power versus those without power; wealthy countries versus poor countries; citizens versus guest/transient workers; higher social standing and mobility versus being stuck in abject status quos; affluent neighbourhoods and schools versus ghettos and the remnants of social Darwinism; ad infinitum.

While most of the world is caught up in dealing with the excruciating minutiae and the vexing exasperations of day-to-day life simply to survive, we in academia are in the privileged position to ponder over the bigger questions confronting humanity. Why do inequities exist in the first place? What are their origins? Are the chapters in this book simply attempts at "patching up" things that are in essence atomically broken., i.e., an allopathic attempt of getting rid of symptoms so we don't have to deal with the real objective roots of problems. Another analogy is that of surgical procedures done on an ad-hoc basis to remedy defects that arise as opposed to caring for the well being of the whole and getting to the root of problems. Or are these chapters, well intentioned attempts around the world to present arguments for the necessity to address social inequities via mathematics education, i.e., to give a deeper meaning to the purpose of education. A nihilist would choose the allopathic (surgical) answer whereas the empathetic individual would choose the latter. Most of us find ourselves somewhere in between, in perpetual but necessary tension to solve the bigger problems around us.

The common bond shared by all the authors in this book is the fact that they are pre-dominantly mathematics educators interested in changing the status quo contributing to the continuation of social injustice in different regions of the world. So, I pose again to the reader the question about the real origins of inequity and injustices within educational and societal mechanisms. Some positions are now presented.

The Darwinian explanation suggests that inequity is simply one of the many natural mechanisms that have arisen over the course of our evolution. If we view ourselves as creatures whose sole purpose in life is to survive and to have progeny, then it is evident that the competition for the same natural resources would leave others in the wake. The strictly Darwinian explanation would suggest that certain groups are doomed to perish simply because they are unable to cope with changes occurring in their environment. Unlike other mammals, we tend to hoard natural resources, much more than we can possibly use and at the same time, we also exhibit tendencies towards altruism which are paradoxical and unexplainable in strictly biological terms. In fact, Charles Darwin (1871) in the *Descent of Man*, posed the question whether the phenomenon of moral behaviour in humans could be explained in evolutionary terms, viz., natural selection. The evolution of social systems (religious, ideological, political) of various kinds are not explainable strictly in Darwinian terms. Comte (1972) proposed a stage theory for our social evolution in which humanity moves from a theological stage onto a metaphysical stage onto a "positive" stage. It is too difficult to explain the meaning of the third stage, but simply put, we reject absolutism of all kinds and we strive for knowledge based on rationality.

The present day economic inequity in the world is best illustrated by the fact that many universities in the West have larger budgets than the GNP of many nations in Africa, Asia and South America. Despite the current state of affairs we are also creatures of ideas who over the course of our evolution have moved away from a strictly clannish and genotypic connection to a memetic[2] connection. We conglomerate over common ideas or ideals as evidenced in the spread of the numerous great world religions, which link together people across a spectrum of class, culture, race, socioeconomic status and nationality. This very book is a memetic product. Similarly ideologies such as Marxism connect people from diverse socioeconomic and cultural backgrounds. Even the so-called phenomenon of "globalization" is nothing new from the point of view of history. There is sufficient historical evidence that even in periods when means of transport and communication had not been developed, oriental civilization penetrated into the West. Iran and Greece were in contact with each other, and many Indians found their way to Greece and vice-versa through this contact (Radhakrishnan, 1964). Asoka's[3] missions to the West, and Alexander's influence on Egypt, Iran, and North West India, produced a cross-fertilization of cultures.

Another big, intensive, but relatively "localized" process, which we may, also call "globalization", occurred in Europe, in the expansion of Christianity in the Middle Ages, in the shadow of the Roman Empire. In the late Middle Ages, States began to take shape as components of a new form of Empire. The scenario resulting from this process of European "globalization", prevails until now. In the sort of jig-saw puzzle which characterize the political dynamics present in this process, the idea of a Nation became strong. States and Nations are different concepts, as well as Political Dynamics and Cultural Dynamics. The political dimension of this process prevailed and something vaguely called State/Nation began to take shape as the primary unit of the European scenario. The Empire which emerged in the Late Middle Ages and the Renaissance as the assemblage of such State/Nations, although fragile, mainly due to power struggle, favored the development of the ideological, intellectual and material bases for building up the magnificent structure of Science and Technology, anchored in Mathematics, supporting a capitalistic socio-economic structure. The expanding capitalism, supported by religious ideology and a strong Science and Technology, had, as a consequence, a new form of globalization, now effectively engaging the entire Globe. The great navigations and the consequent conquest and colonization, completely disclosed the fragility of a possible European Empire. The internal contradictions of State, as a political arrangement, and of Nation, as a cultural arrangement, emerged, in many forms (Sriraman & Törner, 2007).

Religious and linguistic conflicts, even genocide, within a State/Nation became not rare facts. Indeed, they are not over. As a result of all these processes, Education was, probably, the most affected institution. Educational proposals, even curricula, are noticed in this era. The influence of national characteristics interfered with objectives derived from the new World scenario. The development of Science and Technology, obviously related to the educational systems, was unequal. Interchanges intensified. The Industrial Revolution made Science and Technology a determinant of progress. Hence, the enormous competition among European States, which intensified during the 19th century and early 20th century, raised Science and Technology, which became increasingly dependent on Mathematics, to top priority. One terrible consequence of this competition between European states was the advent of colonization, the consequences of which the world is still very much experiencing.

Although many countries in Asia, Africa and South America became "free" from the yoke of colonialism in the last century, this freedom left in its wake uprooted peoples when colonial masters started drawing lines on maps to "equitably" partition land in various regions of the world. Hopefully the reader realizes the irony in my previous statement. There was considerable loss of subsistence lifestyles, loss of indigenous cultures and traditional knowledge. The consequences of colonization were not any different in North America and in Australasia. The outcome of the colonial period of our history was Education as an Institution and a new economic structure being implanted in various regions of the world with the explicit purpose of perpetuating the very structures created to maintain colonialism, namely oppression of the many by a few. Indeed Karl Marx and Friedrich Engels' monumental writings[4] address issues such as exploitation of workers within a capitalistic economic system and the problem of materialism confronting humanity, which would inevitably lead to class struggles and revolutions. Many of the foundational writings of social justice can be traced back to the ideas proposed by Marx and Engels. Today's study of the ecological footprints left by the industrialized nations reveals the obscene differences in resource consumption[5] between rich and poor nations, a natural consequence of materialism run amok as predicted by Marx and Engels.

Paolo Freire (1921–1997), the Brazilian educator and social reformist, came of humble backgrounds. His book *Pedagogy of the Oppressed* (Freire, 1998) is perhaps the most frequently cited Marxist-influenced[6] work in educational literature. Freire (1998) addressed the power dynamics between the oppressed and the oppressors (including the dynamic between teacher and student), and that the way toward liberation is through political movements and political struggle, of which literacy is but one part. Thus his emphasis on *writing*[7] the world, is beyond literacy. Clearly, literacy (i.e., reading the world) is also an integral and necessary part of this process. Freire's banking concept holds that students are knowledgeable beings with the intrinsic capacity of creating knowledge *with the teacher*, as opposed to being empty buckets of ignorance or simply "files" or automatons dependent on the teacher's absolute authority to learn and construct new knowledge. It is also important to note that Freire emphasized critical literacy as opposed to functional literacy. The Organization for Economic Co-Operation and Development (OECD, 2004) defines mathematical literacy as an individual's capacity to identify and understand the role that mathematics plays in the world. Further literacy involves making well-founded judgments and using and engaging with mathematics in ways that meet the needs of each individual's life as a constructive, concerned and reflective citizen. It should be noted that countries like Brazil, China and India are not a part of OECD but are key players in globalization with large vulnerable populations. The essential question is: Does the OECD represent only the interests of the citizens of developed and wealthy countries who are its members or does it also take into consideration the need for equitable and sustainable development with non-members, and more importantly create an awareness of this inequity to students in countries which participate in the PISA.[8] In spite of the good intentions of the OECD, is the push for mathematical literacy around the world simply another mechanism at propagating functionality in the masses as opposed to critical thought and liberation? For instance do large scale tests like PISA include problem solving and problem posing items which make students quantitatively and qualitatively analyse (1) trends in immigration data within OECD and between OECD and non-OECD countries, the causes and consequences thereof; (2) reported incidences of hate crimes against minorities and immigrants in OECD countries, the causes and consequences thereof; (3) comparative data on resource consumption between OECD and non-OECD countries, the reasons for huge discrepancies and their consequences; and (4) data revealing trade deficits and surpluses between OECD and non-OECD countries, the causes and consequences thereof.

Freire (1998) suggested that pedagogical practices should support education for liberation and emphasized problem-posing pedagogies that strive "for the emergence of consciousness and critical intervention in reality" (p.62). Problem posing pedagogies are necessary if the goal of education is to challenge inequities. Freire's writing suggests a pedagogy which promotes greater social awareness or a social consciousness appropriate for initiating major shifts in thinking. An outstanding example of this pedagogy in practice is Gutstein's (2006) work *Reading and Writing the World with Mathematics*. Gutstein's work also points out the obstacles to such a pedagogy within a school system, particularly institutional resistance from administration and other stake holders within a school district.

A nihilist again poses the question: Can emancipatory and social justice pedagogies really free individuals from oppression at a societal level? How can this be possible without it occurring at the individual level first? Freire (1998) himself wrote that the central problem was "How can the oppressed, as divided, unauthentic beings, participate in developing the pedagogy of their liberation? Only as they discover themselves to be "hosts" of the oppressor can they contribute to the midwifery of their liberating pedagogy." Clearly Freire is stating that the oppressed adhere to the oppressor and have to break free. If individuals do not subjectively and intrinsically feel free, how can any educational or social mechanism make this happen no matter how good the intention? Cho & Lewis (2005) recently re-emphasized the aforementioned essence of Freire's pedagogy from the point of view of psychology and the problems with the attempts by Marxist theorists to transform Freire's "pedagogy of the oppressed" into a "pedagogy of revolution". They write that "oppression has an existence in the unconscious such that those that are oppressed form passionate attachments to the forms of power that oppress them" (p.313), and it is necessary for social justice researchers and Marxist theorizers to recognize and address this important issue. Cho & Lewis (2005) formulate several challenges[9] to Marxist theorizers as follows:

> ...part of the discomfort with "revolutionary pedagogy," is that the project of liberation often appears to be presupposing universal notions of what it means to be oppressed, liberated, and how this movement is to be made- often the problem lies in Freire's emphasis on material relations and not on the issue of patriarchy or colonization.... [w]ith no clear resolution to the issue of authority, libratory pedagogies can portray particularist notions of oppression and liberation in universal was and to impose these visions of oppression and liberation upon others through a kind of vanguardism, which can ironically replicate relations of oppression other than overcome them thus returning us to the problem with which Freire begins his analysis in the first place. (p. 314)

a. three shapes, each one with one of three colours

b. two shapes, each one in two sizes as well as in two colours.

The yellow circle (the small one in the second example) could in each case be the neutral block.

The adding rules will be similar to the ones we have already had. In the first example yellow and circle will be weak properties, blue and square will be middle properties, red and triangle will be strong properties. In the second example small, yellow and circle will be weak, big, red and triangle will be strong properties.

Adding a weak property will not change the strength of the property to which it is added. In the first example

middle + middle = strong, strong + strong = middle,
strong + middle = middle + strong = weak.

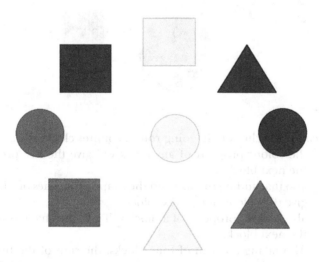

Figure 2.6

Here is the circle of the first example (Figure 2.6). Putting the neutral in the centre of the circle we can verify that going round counter clockwise any two neighbouring blocks will add up to the one immediately after the second block.

It also easily seen that

the swapped colour property plus the shape property of a block gives the colour property of the next block, and that the swapped colour property plus the swapped shape property of any block will be the shape property of the next block.

The circle of the second example will look like Figure 2.7.

Figure 2.7

You can easily check that, going round counter clockwise
the colour property of any block will give the size property of the next block,
also that adding the size and the shape properties of a block will give the colour of the next block
also the size property of a block will give the shape property of the next block.
Also taking any run of four blocks, the sum of the first two is always the fourth.

It is also very easy to see how the circles can be constructed by juxtaposing sequences where 0 stands for weak, 1 for middle and 2 for strong. This is what we get from the first example:

$$2 \quad 1 \quad 0 \quad 1 \quad 1 \quad 2 \quad 0 \quad 2 \quad 2 \quad 1 \quad 0 \quad 1 \quad 1$$
$$0 \quad 1 \quad 1 \quad 2 \quad 0 \quad 2 \quad 2 \quad 1 \quad 0 \quad 1 \quad 1 \quad 2 \quad 0,$$

the sequence being 1 1 2 0 2 2 1 0

This is what we get from the second example:

```
S   S   S   W   W   S   W   S   S   S   W   W   S
S   S   W   W   S   W   S   S   S   W   W   S   W
W   S   S   S   W   W   S   W   S   S   S   W   W
```

here the repeating sequence is S S S W W S W

If you want to get to the paddling pool, you can look at two variables with two values each. Let us take yellow and red circles and squares. Our "circle" will now look like Figure 2.8:

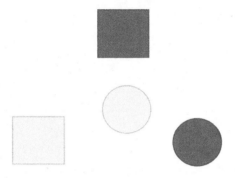

Figure 2.8

Any two blocks will add up to the third block and going round the circle clockwise

> strength of shape + strength of colour = strength of shape next
> strength of shape   = strength of colour next

and the juxtaposed cycles will look like this:

```
S   S   W   S   S   W   S   S   W
S   W   S   S   W   S   S   W   S,
```

If you know something about the mathematics learning theory I have developed, you will remember the principle of "throwing them in at the deep end" Then afterwards swimming at the shallow end will be just a piece of cake! Of course don't make it too deep, or your customer might drown! In the series of examples I have given, the second example is "deeper" than the first! So I have been a little bit careful!

CHAPTER 3

LOGIC BLOCKS TO OTHER EMBODIMENTS

Zoltan Paul Dienes

SENTENCES AND NUMBERS

Write the following sentences, each one on a separate small piece of cardboard:

John often plays inside,
John often works inside,
John seldom plays inside,
John seldom works inside,
Mary often plays inside,
Mary often works inside,
Mary seldom plays inside,
Mary seldom works inside,

John often plays outside,
John often works outside,
John seldom plays outside,
John seldom works outside,
Mary often plays outside,
Mary often works outside,
Mary seldom plays outside,
Mary seldom works outside.

Then write the following numbers, each on a small piece of cardboard:

1	2	3	6	5	10	15	30
7	14	21	42	35	70	105	210

You will have noticed that the above numbers are obtained, starting with the number 1, by multiplications by 2 or by 3 or by 5 or by 7 or by several

Mathematics Education and the Legacy of Zoltan Paul Dienes, pages 37–48

such multiplications. So each of the numbers either does or does not have 2 as a factor, 3 as a factor, 5 as a factor, 7 as a factor. I shall call multiples of 2 the 2-numbers (even numbers), those that are not multiples of 2 (odd numbers) the non-2 numbers. In a similar way I shall talk about 3-numbers, 5-numbers and 7-numbers and of course about non-3 numbers, non-5 numbers, and non-7 numbers.

THE "SUM" OF TWO SENTENCES AND OF TWO NUMBERS

Each sentence consists of four parts, each part being either "weak" or "strong." It is quite arbitrary which part is strong and which is weak, but I shall make the following choices:

| Mary, | often, | works, | outside | will be strong parts |
| John, | seldom, | plays, | inside | will be weak parts |

To "add" two sentences the corresponding parts are "added" according to these rules:

weak + weak = weak,   weak + strong = strong
strong + weak = strong,   strong + strong = weak

The "sum" of two numbers will not be obtained by adding the numbers but by multiplying them. When you have multiplied any two of your numbers, look out for any factor that occurs twice in your product. Just leave this factor out. The resulting number will be your "sum". For example

$30 \times 14 = 2 \times 3 \times 5 \times 2 \times 7$, so leaving out the 2's we get 105.

It would look very bad to write $30 \times 14 = 105$ (since this is not true) and even worse to write $30 + 14 = 105$, we could use another symbol such as the word "with" and write

$$30 \text{ with } 14 = 105$$

MAKING A CIRCLE OF SENTENCES

Arrange your sentences in a circle, leaving out the weakest sentence, namely

John seldom plays inside

so that John sentences and Mary sentences come in this circular order:

Mary Mary Mary Mary John John John Mary John John Mary
Mary John Mary John

Now, exchanging Mary sentences for Mary sentences and John sentences for John sentences make sure "often" and "seldom" also come in that order, namely as

often often often often seldom seldom seldom often seldom seldom
often often seldom often seldom

Now again, exchanging sentences in which the first two parts are the same, make sure that "works" and "plays" come in the above order, namely as

works works works works plays plays plays works plays plays works
works plays works plays

Then do the same for outside and inside, exchanging sentences in which the first three parts are the same. You will need to have:

outside outside outside outside inside inside inside outside inside
inside outside outside inside outside inside

Now look at any five consecutive sentences along your circle. Find the sum of the first and the second sentence of your run of five. Do you get the fifth sentence of your run?

Try with several runs of five. Do the first and the second always "add up" to the fifth?

MAKING A CIRCLE OF NUMBERS

Leaving out the number 1, try to arrange the fifteen numbers so that even and odd numbers are arranged in the following order:

even even even even odd odd odd even odd odd even even
odd even odd

Then, exchanging even with even and odd with odd, make sure that the 3-numbers and the non-3 numbers follow the same pattern, namely:

3   3   3   3   N3   N3   N3   3   N3   N3   3   3   N3   3   N3

Now, exchanging even 3's with even 3's, odd 3's with odd 3's, even N3's with even N3's and odd N3's with odd N3's make sure that the 5's and N5's also follow the same pattern. Similarly arrange the 7's and N7's, but without spoiling the arrangements of the 2's, 3's and 5's.

Now take five consecutive numbers of your circle of numbers.
Is it true that
(first number) with (second number) = (fifth number)?

Can you make any particular number correspond to one particular sentence?

Write out the rules that let you pass from numbers to sentences and vice versa.

Here is one possible number sequence:

2 3 5 7 6 15 35 42 10 21 30 105 210 70 14

It is clear that 2 with 3 = 6, 3 with 5 = 15, 5 with 7 = 35, 7 with 6 = 42, 6 with 15 = 10 leaving out the factor 3 in the last "with".

RULES FOR THE NEXT ONE ROUND THE CIRCLE

The following fifteen sentences might have come out if you had followed the instructions:

Mary seldom plays inside
John often plays inside
John seldom works inside
John seldom plays outside
Mary often plays inside
John often works inside
John seldom works outside
Mary often plays outside
Mary seldom works inside
John often plays outside
Mary often works inside
John often works outside
Mary often works outside
Mary seldom works outside
Mary seldom plays outside

Let us find a way of telling what the next sentence is that comes just after the one we are looking at. It is easy to see that there is always a Mary after an outside and a John after an inside. A John outside or a Mary inside is followed by an often and a John inside or a Mary outside is followed by a seldom. An often is followed by works and a seldom is followed by a plays. Finally a plays is followed by an inside and a works by an outside.

Try to find rules for getting the sentence just before a given sentence or you might try for the last but one or the next but one.

What would be the rules for finding the next number in this sequence?

$$2 \quad 3 \quad 5 \quad 7 \quad 6 \quad 15 \quad 35 \quad 42 \quad 10 \quad 21 \quad 30 \quad 105 \quad 210 \quad 70 \quad 14$$

It is not hard to see that any number that has a factor 7 is followed by an even number and those not having a factor 7 are always followed by an odd number. If a number has just one of the factors 2 or 7, then a number with a factor 3 follows. If a number has both or neither of these factors, then the next number will not have the factor 3. A number with a factor 3 will always be followed by a number with factor 5. A number that does not have the factor 3 will be followed by a number that does not have the factor 5. Numbers with the factor 5 are always followed by a number with factor 7. If a number does not have 5 as a factor, the next number will not have 7 as a factor.

EXTENDING THE CIRCLE

New Adding Rules and Making a Circle of Numbers

Let us start with numbers this time. Let us think of a number that has not only factors, but also factors squared. A good number would be 900, since

$$900 = 2 \times 2 \times 3 \times 3 \times 5 \times 5$$

Let me first define the operation of "swapping". To "swap" a number, we alter any factor that is present to the second power, to just the first power, and if the factor is present just to the first power, "swapping" will mean that this factor will now be taken to the second power.

for example: swap $(2 \times 2 \times 3 \times 5) = 2 \times 3 \times 3 \times 5 \times 5$.

Since we now have three possibilities for any factor, namely: non-occurrence, occurrence to the first power, occurrence to the second power, any prime factor occurring three times after a with-ing will have to be deleted in the "with" of the two numbers.

Let us start our circle with the three numbers: 5, 3, 2. Let us swap the first one, and with-it to the swapped second one, which we can with to the third one. So our fourth number will be

$$5 \times 5 \times 3 \times 3 \times 2 = 450$$

So our sequence now looks like this:

$$5 \quad 3 \quad 2 \quad 450$$

Now swapping the second and the third and putting their "sum" with the fourth since $3 \times 3 \times 2 \times 2 \times 450 = 3 \times 3 \times 2 \times 2 \times 5 \times 5 \times 3 \times 3 \times 2$ so we get

$$5 \times 5 \times 3 = 75$$

having deleted three 2's and three 3's. So our sequence now will look like this:

$$5 \quad 3 \quad 2 \quad 450 \quad 75$$

Now swapping the 2 and the 450 and putting it with 75, we have

$$2 \times 2 \times 5 \times 3 \times 2 \times 2 \times 3 \times 5 \times 5$$

we get

$$3 \times 3 \times 2 = 18$$

after deleting three 2's and three 5's, so we now have

$$5 \quad 3 \quad 2 \quad 450 \quad 75 \quad 18$$

Carrying on like this we eventually obtain the sequence

5	3	2	450	75	18	225	36	30	100	20	180	90
25	9	4	60	45	12	15	6	900	10	50	150	300

Let us give the factors a weight. The number can be weak from the point of view of a prime factor if that factor is not a factor of that number. If the factor is present to the first power, its strength will be called middle. If the factor is present to the second power, its strength will be called strong. I shall use the letters W, M and S for weak, middle and strong respectively. I shall refer to the factors in the order factor 2 or 4   factor 3 or 9,   factor 5 or 25

For example 60 will be referred to as SMM since $60 = 2 \times 2 \times 3 \times 5$.

Let us give each number some points. One point is awarded for a factor that occurs just to the first power, two points for a factor that occurs to the second power. So middle means one point, strong means two points. for example 900 will have 6 points, as it has three squared factors, or $10 = 2 \times 5$ has 2 points, as it has two factors, each to the first power.

We can now establish some rules for getting the third number that just comes after two consecutive numbers in our circle. The symbol (1) will mean our first number, the symbol (2) our second number. The numbers before the brackets refer to the factors 2 or 3 or 5. A "score" can be defined as follows:

$$\text{score} = 2(1) \quad + \quad 3(1) \quad + \quad 3(2) \quad + \quad 2 \times 3(1) \quad + \quad 2 \times 3(2)$$

but don't forget to add them in modulo 3. This means that 3 counts as 0, 4 as 1, 5 as 2, 6 as zero and so on.

If score = 0 there is no 5 in the third number

If score = 1 there is a factor 25 in the third number

If score = 2 there is a factor 5 but not a 25 in the third number.

If one is a no2 and the other a 2 or each one is a 4 then third has a 9

If one is a 4 and the other a 2 or each one is a no2 then the third has no 3

If one is a 4 and the other a no2 or each one is a 2, then the third has a 3

Now let the "score" be the sum of all the points, arising from both your first number and your second number, of course added up in modulo 3. Then

score = 0   means your third number is a no2

score = 1   means your third number has a factor 4

score = 2   means your third number has a factor 2 but not 4.

Sentences with Three Values for Each Part

Let us take

John,	Mary,	The dog	as values for the first part,
is always,	is sometimes,	is never	as values for the second part

working,	playing,	sleeping	as values for the third part
John,	is always,	working	will be called strong
Mary,	is sometimes,	playing	will be called middle
The dog,	is never,	sleeping	will be called weak.

So the sentence "The dog is never sleeping" will be the neutral sentence, it being the all-weak sentence. The same adding rules as before will be valid for adding sentences, not forgetting that

$$\text{strong + strong = middle, middle + middle = strong,}$$
$$\text{strong + middle = weak}$$

and adding a weak will not alter the strength, as before.

Swapping will again be as before,

$$\text{swap (weak) = weak, swap (strong) = middle, swap (middle) = strong}$$

Let us start with the three sentences:

The dog is never playing

The dog is sometimes sleeping

Mary is never sleeping.

Let us use the following rule for the construction of our circle of sentences:

$$\text{swap (first sentence) + swap (second sentence) + third sentence =}$$
$$\text{fourth sentence}$$

by means of which we shall obtain the following sequence:

The dog is never playing

The dog is sometimes sleeping

Mary is never sleeping.

Mary is always working c

The dog is sometimes working c

Mary is always sleeping c

The dog is always working c

John is always sleeping c

Mary is sometimes playing c

John is never working c

John is never playing c

John is always playing c

Mary is always playing c

Here is the second half of our sequence, which is the swapped version of the first half.

The dog is never working

The dog is always sleeping

John is never sleeping

John is sometimes playing

The dog is always playing

John is sometimes sleeping

The dog is sometimes playing

Mary is sometimes sleeping

John is always working

Mary is never playing

Mary is never working

Mary is sometimes working

John is sometimes working

and then we start again with the first sentence of the first half.

Now let us see what the succession rules are in our sequence, Will they be similar to the rules we had in our number sequence? Let us see.

The easiest rule to spot is about telling which verb comes in the next sentence, You will easily check these simple rules:

After a dog sentence there is always a sleeping sentence

After a Mary sentence there is always a working sentence

After a John sentence there is always a playing sentence.

To tell whether the next sentence is a John one or a Mary one or a dog one, we have the following rules:

dog never or Mary always or John sometimes → dog sentence next

dog sometimes or Mary never or John always → Mary sentence next

dog always or Mary sometimes or John never → John sentence next

To tell whether the next sentence is with never or with sometimes or with always, we have the following rules:

dog sleeping or Mary playing or John working → never sentence next

dog playing or Mary working or John sleeping → sometimes sentence comes next

dog working or Mary sleeping or John playing → always sentence comes next

We could shorten our sentences to two-part sentences, and have either 3 values or maybe even 5 values for each part. Let us start with three values. Our sentences could be

The dog is sleeping,	Mary is sleeping	John is sleeping
The dog is playing	Mary is playing	John is playing
The dog is working	Mary is working	John is working

The dog is sleeping will be our neutral sentence, if we remain by the same strength values that we had before. If we start with two sentences and obtain the third one by adding the first to the second, we can again proceed to reach all our non neutral sentences, thus completing a circle of eight sentences. The circle could, for example, be the following:

The dog is playing

Mary is sleeping　　　　　　Mary is working

Mary is playing　　　　　　　　　John is working

John is playing　　　　　John is sleeping

The dog is working

where, moving counter clockwise, two successive sentences always add up to the third sentence following the two. It is also easy to see that if we add the strength of the first part to the strength of the second part of a sentence, we obtain the strength of the first part of the next sentence.

Also the strength of the first part of any sentence is always equal to the strength of the second part of the next sentence.

A corresponding number sequence would be:

$$2, \quad 3, \quad 6, \quad 18, \quad 4, \quad 9, \quad 36 \quad 12, \quad \text{back to 2 etc.}$$

If you multiply any two consecutive numbered and delete any factor occurring three times, you will always get the next number in the sequence. The sequence can be written using strength values as follows, the first value showing the 2-strength, the second value the 3-strength:

$$MW, \quad WM, \quad MM, \quad MS, \quad SW, \quad WS, \quad SS, \quad SM$$

where it is easily verified that adding the strength values of the first and of the second parts of any number will yield the strength of the first part of the next number. Also the strength of the second part of any number is equal to the strength of the first part of the next number.

What about trying 5 values for each of two parts of our sentences. For example we could have

> For the first part:
> The king, the queen, John, Mary, the dog

> For the second we could have:
> is working, is playing, is eating, is resting, is sleeping

with strength values of 4, 3, 2, 1, and 0 respectively, in the above order. We can refer to the sentences by stating the strength values, so for example

"The queen is eating" would be referred to as 32.

We can start with the two sentences 10 and 01 and use the generation rule, using modulo 5 arithmetic:

$2 \times$ (first sentence) $+ 3 \times$ (second sentence) $=$ third sentence.

We shall obtain the following sequence:

| 10 | 01 | 23 | 11 | 24 | 34 | 30 | 03 | 14 | 33 | 12 | 42 |
| 40 | 04 | 32 | 44 | 31 | 21 | 20 | 02 | 41 | 22 | 43 | 13 |

The next sentence coming after any given sentence is given by the rule:

$2 \times$ verb $\qquad =$ strength of noun of next sentence
noun $+ 3 \times$ verb $=$ strength of verb of next sentence

The cyclic sequence followed by both noun and verb is easily detectable from the above given sequence of sentences. The noun sequence begins at the eighth member of the verb sequence.

We could now see what happens if we try three variables, each variable having just two values.

We could come back to John and Mary, either working or sleeping, inside or outside. We shall have the sentences:

Mary is sleeping inside,	Mary is sleeping outside,
Mary is working inside,	Mary is working outside
John is sleeping inside,	John is sleeping outside,
John is working inside,	John is working outside

or alternatively the numbers:

$$1, \quad 2, \quad 3, \quad 6, \quad 5, \quad 10, \quad 15, \quad 30$$

expressed in terms of "strengths" these would be written:

WWW SWW WSW SSW WWS SWS WSS SSS

We can make a sequence out of these numbers or sentences as follows:

SSS SWS WWS WSW SWW WSS SSW

The strength of the second part of a sentence will always give the strength of the first part of the next sentence. Adding the strengths of the first and of the third part of any sentence will give the strength of the second part of the next sentence.

The strength of the first part of a sentence will give the strength of the third part of the next sentence. I will leave it to you to formulate the succession rules for the number sequence.

CHAPTER 4

SOME REFLECTIONS ON ORDER AND DENSITY

A Child's Path to the Bolzano– Weierstrass Theorem

Zoltan Paul Dienes

THE SUBDIVISION STORY

I shall take a story-line as a starting point of my suggestions.

Let us suppose that you want to go and visit a friend. The road to this friends is very long and leads through a forest. You start walking along this road and of course you get tired. You remember that somewhere in the forest lives a game-keeper and you stop there for a rest. The gamekeeper gives you a nice hot cup of tea, so soon you depart, quite refreshed. Eventually you reach your friend's house. On your way back home, before you get to the gamekeeper's house, you already feel tired and you sit down in the shade of a large oak tree.

Having rested for a bit, you proceed to the gamekeeper's cottage. He invites you in again for a cup of tea, which you gratefully accept. When you feel rested you say your thanks and take your leave. On your way home there is

Mathematics Education and the Legacy of Zoltan Paul Dienes, pages 49–66
Copyright © 2008 by Information Age Publishing

another large oak tree in the shade of which you rest for a while. Then you continue along the road and you eventually reach your home.

We have now "subdivided" the road to your friend's house like this:

Your house —— oak tree——gamekeeper's——oak tree ———friend's
   house   house

The next day you want to go again to see your friend, but you've had a bad night and so you are more tired. So you don't even make it to the first oak tree but stop in field and pick some daisies. You go on and have a rest at the large oak tree. On the way to the gamekeeper's house you add some more daisies to your bunch and you accept a cup of tea at the gamekeeper's house, after which, along the road you pick more daisies before getting to the second oak tree where you rest again. You add some more daisies to your bunch from a field on the way to your friend's house and finally you reach your friend's house. He puts your bunch of daisies in water and gives you a good meal.

We have now subdivided the road in this way:

home daisies oak tree daisies gamekeeper daisies oak tree
daisies friend

Theoretically we could go on indefinitely subdividing the road between your house and your friend's house by interspersing more and more activities which would mark the positions of subdivision.

There are a lot of interesting birds in the forest so you could decide to do a bit of bird watching. So you would do this before you reached the first daisy field, then again after this daisy field but before the oak tree. Then you would watch birds again between the oak tree and the next daisy field and again between this daisy field and the gamekeeper's cottage. Then there would be more bird watching before the next daisy field, and again between this daisy field and the oak tree, then again between the oak tree and the next daisy field and finally between this daisy field and your friend's house. So now your journey is punctuated by the following events:

Home birds daisies birds tree birds daisies birds tea birds daisies
birds tree birds daisies birds friend

We have already split the road into sixteen successive sections, and of course we could go on inventing more activities and split the road into more and more sections. For example we could take an apple with us and take a bite of the apple somewhere in each of the sixteen sections. We would then have the following sequence of activities:

The first bite of the apple
The first bird watching
The second bite of the apple
The first daisy field
The third bite of the apple
The second bird watching
The fourth bite of the apple
You get to your first tree
You take the fifth bite of the apple
The third bird watching
You take the sixth bite of the apple
You reach your second daisy field
You take the seventh bite of the apple
You reach the third bird watching spot
You take the eighth bite of the apple
You get to the gamekeeper's house
You take the ninth bite of the apple
You do the fifth bird watching
You take the tenth bite of the apple
You pick daisies in the third daisy field
You take the eleventh bite of the apple
You enjoy the sixth bird watching
You take the twelfth bite of the apple
You get to your second tree
You take the thirteenth bite of the apple
You do your seventh bird watching
You take the fourteenth bite of the apple
You reach your fourth daisy field
You take the fifteenth bite of the apple
You reach the eighth bird watching spot
You take the sixteenth bite of the apple
You get to your friend's house

I leave it to you to invent further activities, thus splitting the road into more and more parts, but before going any further, it might be a good idea to investigate how we can tell just where you are on your journey to your friend's house, which opens up the problem of how to code the various positions.

The diagram follows (Figure 4.1).

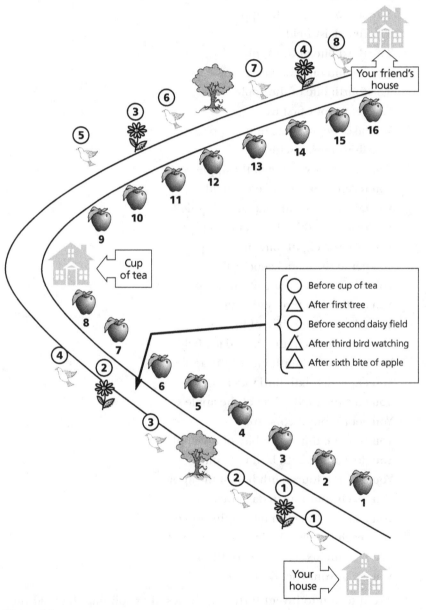

Figure 4.1

THE CODING PROBLEM

Let us use a circle to indicate that you have not passed a particular activity and use a triangle to show that you have passed it. So just one circle will mean that you have not reached the gamekeeper's house where you have a cup of tea, and just one triangle will mean that you have had your tea and have left the gamekeeper's cottage. If we use two symbols, the first one will refer to the gamekeeper's house and the second one to the oak trees. So we shall have:

Before the first oak tree	After the first oak tree but before your tea	After your tea but before the second oak tree	After the second oak tree
○ ○	○ △	△ ○	△ △

Now we can bring in the third symbol, which will tell us whether you have not yet reached a daisy field or whether you have just picked some daisies, a circle again indicating that you are not there yet and a triangle that you have just been there. Here are the codes that use three symbols:

○ ○ ○	○ ○ △	○ △ ○	○ △ △	△ ○ ○	△ ○ △	△ △ ○	△ △ △
Before first daisy	After 1st daisy but before 1st tree	After 1st tree but before 2nd daisy	After 1st tree after 2nd daisy	After tea before 2nd tree before 3rd daisy	After tea before 2nd tree after 3rd daisy	After tea after 2nd tree before 4th daisy	After tea after 2nd tree after 4th daisy

The bird watching situation will be represented by the fourth symbol, not having started the bird watching being represented by a circle and having done it by a triangle. Here is the sequence with the bird watching situations encoded in the fourth place:

○○○○ ○○○△ ○○△○ ○○△△ ○△○○ ○△○△ ○△△○ ○△△△
△○○○ △○○△ △○△○ △○△△ △△○○ △△○△ △△△○ △△△△

The biting of your apple will be encoded in the fifth place, again no biting being encoded by means of a circle and having taken the bite by means of a triangle. So, for example this code: △ △ ○ △ △ will mean that you've had your cup of tea and you've passed the second oak tree but you have not yet reached the next daisy field but you have passed the next bird watching spot after which you have taken the bite of your apple that was due then. Or the code: ○ △ ○ △ △ will mean that you have not reached the gamekeeper's cottage but you've passed the first oak tree but you have not reached the next daisy field but you've been past the next bird watching spot and you've taken the bite of your apple due after the last bird watching. So a code will tell us which part of the road we are talking about although we do not know how "far" we have come. The gamekeeper's cottage could be much nearer your house than your friend's house, all we know is that it is between your house and your friend's house. The same applies to all the other spots defined by our activities. We can only talk about "before" and "after" and not about "how far?"

OUT OF TWO WALKERS, WHICH ONE IS IN FRONT OF THE OTHER?

Let us imagine that two of you walk from your house to your friend's house. You start at different times, so at any given moment one will be in one section of the road, the other will be in another section. There must be a way of telling which section comes before which. In fact you have probably worked out the rule, which is given below (Figure 4.2):

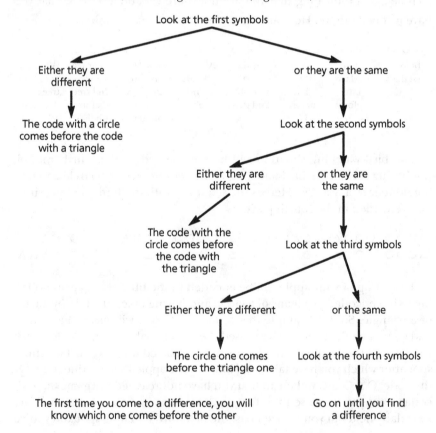

Using the above given rule, it should be quite easy to decide, given two codes, which one symbolizes a walker who is ahead of the other walker. For example let us look at the two codes:

○ ○ ○ ○ △   ○ ○ △ △ ○

The first symbols are equal (they are both before the gamekeeper's house), also the second symbols are equal (They are both before the first oak tree). The third symbols are different. The second one has already picked some daisies but the first has not yet reached that daisy field, so the walker represented by the second code is in front of the one represented by the first code,

Children might take quite a while to figure out this rule, as often they will tell you that the one with more triangles must be in front. In the above example the comparison of the number of triangles in each code would give the wrong answer to the question of which one is ahead of the other. It is always best to keep a track in front of the children, so that they can work out where each walker is, as given by the codes, so that they can realize their error and come to realize the correct rule for deciding which one is in front.

We could think of a road on which we think of each section in three instead of two parts. For example there could be two houses between your house and your friend's house, which would give us three sections: from your house to the first house along the road, then from the first house to the second house, and finally from the second house to your friend's house, after which we can think of two oak trees along each of these sections, which would create a total of nine sections. Along each of these nine sections there would be two daisy fields, thus creating 27 sections. Within each of these sections there would be two bird watching spots, thus yielding 81 sections. Theoretically we could continue indefinitely, inventing more activities and thus creating more sections.

For the cover story we might say that you get tea in the first house and coffee in the second one. In this way we are less likely to mix up the houses. The first oak tree could be small, the second one could be big. For the bunch of flowers you give to your friend, the first field could be full of daisies, the second one full of buttercups. For the bird watching you could first watch doves and then eagles. Finally you could have an apple and a pear, biting the apple first and the pear at the next stop. And you could go on as long you feel like it and invent more such "double" activities.

To codify the above we would need three symbols. Each section would be split into three parts, the first part, the middle part and the last part. The first part could be symbolized by means of a circle, the middle part by means of a triangle and the last part by means of a square. I leave it to you to write out the rule for telling who is before whom, using circles, triangles and squares.

If we put nine activities into each section that we have already constructed, thus making ten sub-sections into which each section is split, we can use
the symbols

$$0, 1, 2, 3, 4, 5, 6, 7, 8, 9$$

and we shall have the usual decimal notation. Of course it would be much
more difficult if not downright impossible even to attempt to draw a picture
of such a road! So I suggest that we stay with a two-way subdivision!

The diagram (Figure 4.3) for subdivision in three parts follows.

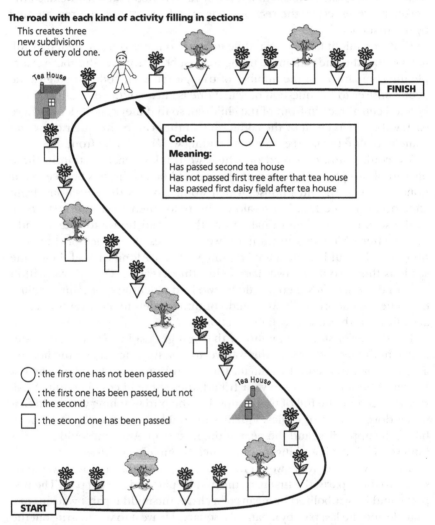

The road with each kind of activity filling in sections

This creates three new subdivisions out of every old one.

Tea House

FINISH

Code: ☐ ○ △

Meaning:

Has passed second tea house
Has not passed first tree after that tea house
Has passed first daisy field after tea house

○ : the first one has not been passed

△ : the first one has been passed, but not the second

☐ : the second one has been passed

Tea House

START

Figure 4.3

THE "IN BETWEEN" PROBLEM

Think of two positions of which you know the code. Let us call them A and B where B is further along the track than A. Another position X is said to be between A and B if X is after A but is before B. Is it always possible to find a position X between any two positions?

For example, taking the two positions

A = △ ○ ○ △ △   B = △ ○ △ △ ○
there is a section coded by   X = △ ○ △ ○ ○

which is after the first one but before the second one. All three have had their cups of tea and all three are before the second oak tree. X has picked some daisies, but A has not yet reached the daisy field so it must be before X. Both B and X have picked daisies but only B has reached the bird watching, while X has not done so. So B is after X, and we see that X is really between A and B.

Sometimes it is not possible to insert an "in between one" without inventing more activities. We could make the convention that if only five symbols are given, the sixth symbol is always a circle in other words the next activity has not yet been reached. Let the walkers take a grapefruit with them and a taste of that can be our sixth activity.

Let the first person and the second person be at these positions

○ △ ○ △ △ ○   ○ △ △ ○ ○ ○

The last circle in each code means that neither person has reached the grapefruit tasting position.

If we wish to place somebody between these two, we must place him at

○ △ ○ △ △ △

which has had his grapefruit whereas the first person has not, so this proposed "in between one" is indeed beyond the first person. The second person is beyond the "in between one" as he has picked daisies while the "in between one" has not, although he did everything else afterwards!

So he is "just" behind the second person, who is in the next section forward!

Make sure you understand all this by working on an actual track which has been constructed right up to the grapefruit tasting!

You will no doubt see now that we can always put somebody "in between" two people, as long as the people are always in the first part of their sections, and if necessary we might need to invent another activity and so create more subdivisions.

BREAK-THROUGH TO "INFINITY"
AS "YOU CAN ALWAYS DO IT AGAIN"

Now we can take the leap into the "infinite", thus daring an incursion into real mathematics! Let us say that our codes do not have to stop, but can go on indefinitely. So each code will be "infinitely long"! Of course for each such code we shall have to specify some clearly defined rule as to how our circles and triangles will follow one another. At this point, we might also become more "mathematical" and use the symbol for zero instead of a circle and the symbol 1 instead of a triangle.

So the position of a person will be defined by an infinite sequence of zeros and ones, symbolizing an infinite sequence of sections, each section being a part of a previously defined section. Let us make the convention that if we write a code consisting of a finite number of zeros and one's, then it should be understood that all the rest of the code will consist of zeros, meaning that the person symbolized by our code will always opt for the first part of each successive subdivision after the one symbolized by the last written symbol. For example

$$0 \quad 1 \quad 0 \quad 1 \quad 1 \quad 0 \dots\dots\dots\dots\dots\dots\dots\dots\dots$$

will in fact be left standing at the place where he took a bite of his apple and has not started at all towards the next subdivision of the road., meaning that his code would really be

$$0 \quad 1 \quad 0 \quad 1 \quad 1 \quad 0 \quad 0 \quad 0 \quad 0 \quad 0 \quad 0 \quad 0 \quad 0 \quad 0 \quad 0 \quad 0 \quad 0 \quad 0 \quad 0 \quad 0 \quad 0 \dots\dots\text{ad inf.}$$

So to have stopped just where an activity was carried out means that the rest of the code consists of zeros. The awkward thing is that this same position could also be defined by using a lot of one's instead of a lot of zeros. This is because our apple chewer is in the last section of the last section of the last section and so on of the sections of the road that he has just passed whereas he is also in the first section of the first section of the first section and so on of all the sections that he has not started on yet! So there are two ways in which we can define a position at which a person has stopped, having just done an activity!

The two codes would look like this:

(any sequence of zeros and ones) 1 0 0 0 0 0 0 0 0 0 0 0 0 0 ... ad inf.

(the same sequence as above)   0 1 1 1 1 1 1 1 1 1 1 1 1 1 ... ad inf.

DENSITY

The "infinite sequences" of zeros and ones lead to the posing of an interesting problem the solution of which will make things clearer. Let us assume that we have already played the "in between game" and have tried to place "people" in between other "people" on our road. Let us consider the example of Anne and Bill and a person to be inserted in between them:

Anne being at   0 1 0 1 1 0 0 0 0 0 0 0 0   …..

Person in between 0 1 0 1 ? ? ? ?   ……..

and Bill being at   0 1 0 1 0 1 1 1 1 1 1 1 1   …..

and let us try to put a person between Anne and Bill. The children looking at the two codes will certainly say that Anne is further along the road than Bill, as at the fifth place for the first time there is a difference and at this place Anne has a 1 and Bill has a 0. So Anne has taken this particular bite of the apple whereas Bill has not.

Let us call the person to be inserted Charlie. Then Charlie's code must begin with what is common to Anne's and Bill's codes, as shown above. But what is Charlie's fifth symbol? Let us suppose it is 0. Whatever symbols come after this, Charlie will always be not as far as Anne but if the sixth symbol is 0 Charlie will be placed not as far as Bill, therefore the sixth symbol must be 1.

By the same argument the seventh symbol must also be 1 and so on. In fact this means that Charlie and Bill have exactly the same code. So Charlie's fifth symbol cannot be 0. Therefore it must be 1. If we put any 1's after this, Charlie's code would place him ahead of Anne so from the sixth symbol on we must have all 0's in Charlie's code which makes Charlie's code identical with Anne's. So however we try to write Charlie's code, we end up either with Anne's or with Bill's. In fact there is "no room" for Charlie! In fact Anne and Bill are at the very same spot on the road! So any position whose code ends with an "infinite number" of 0's can also be symbolized by means of a code with an "infinite number" of 1's.

So to avoid any complications, we could make the convention that codes ending in an infinite number of 1's are prohibited!

Allowing the above prohibition, we can now say that we can always find a position between any two different positions.

Out of any two codes one always represents a position after the other. We say that the positions are ordered.

Also there is always a code that represents a position between any two positions.

We say that the positions are densely distributed along the road.

Order and density are two of the most fundamental ideas in mathematics.

SEQUENCES OF POSITIONS

The code for the position at the "start" of the road, at which point you have not done any activities of your journey, is

$$0\ 0\ 0\ 0\ 0\ 0\ 0\ 0\ 0\ 0\ 0\ 0\ 0\ 0\ 0\ 0\ 0\ 0\ 0\ \dots\dots \quad \text{ad inf.}$$

and the code of your friend's house, having done all the activities, is

$$1\ 1\ 1\ 1\ 1\ 1\ 1\ 1\ 1\ 1\ 1\ 1\ 1\ 1\ 1\ 1\ 1\ 1\ 1\ \dots\dots \quad \text{ad inf.}$$

which we can call the finish.

All our positions are between the start and the finish. Let us pick any position between start and finish, and let us call this position L. Then let us pick a position before L. Let us call this our first position. Now pick a position between the first position and L. Let us call this our second position. Then pick a position between the second position and L, which will be our third position. Since there is always a between, this process can go on for ever. We say that we can construct an increasing sequence of positions, meaning that each position is "further on" towards the finish than the position before it.

Here is an example of how we can construct such an increasing sequence of positions.

$$\text{Let us say that} \quad L = 1\ 1\ 0\ 0\ 0 \dots\dots\dots\dots$$

We can choose our first position as $\quad 1\ 0\ 0\ 0\ 0 \dots \dots < 1\ 1\ 0\ 0\ 0 \dots\dots$
then the second one as $\quad\quad\quad\quad 1\ 0\ 1\ 0\ 0 \dots\dots < 1\ 1\ 0\ 0\ 0 \dots\dots$
then the third one as $\quad\quad\quad\quad\ \ 1\ 0\ 1\ 1\ 0 \dots\dots < 1\ 1\ 0\ 0\ 0 \dots\dots$
then the fourth one as $\quad\quad\quad\quad 1\ 0\ 1\ 1\ 0 \dots\dots < 1\ 1\ 0\ 0\ 0$

and so on. Each position of our sequence is further towards L than the previous one and each position is before L. We can always construct such an increasing sequence, as long as we never include the finish.

INTERVALS, SETS AND PROPERTIES

An interval consists of all the positions between two given positions. These two positions are called the end-points of the interval. If the endpoints are not included, the interval is called an open interval. If the endpoints are included, the interval is called closed. It is always possible to decide whether any given position is one of the positions of the interval or not. Suppose that the endpoints are A and B, then a position X belongs to the open interval (A, B) if

$$A < X \quad \text{and} \quad X < B$$

If one or both of the above are not true then X does not belong to the open interval (A, B).

Sometimes it is useful to talk about a number of positions, which we lump together thus forming a set of positions. There are two ways of defining a set: one is to enumerate the positions that we wish to consider as belonging to the set, another way is to state a property a position must have in order for it to be a member of the set. Using the first way, we can only form sets with a finite number of members, as we cannot enumerate an infinite number of anything.. We can specify any property we like, as long as it is always possible to decide whether the code of a position does or does not have the property in question. In other words we must be precise in defining a property.

For example we could consider the positions whose codes, after a certain point, consist of a pattern of zeros and ones which repeats indefinitely. For example

0 1 1 0 1 1 0 1 1 0 1 1 0 1 1 0 1 1 0 1 1... and so on,
repeating 0 1 1 indefinitely.

Or the recurring pattern does not have to start at the beginning, for example

0 1 0 0 0 1 0 0 1 0 1 0 1 0 1 0 1 0... repeats 0 1 indefinitely
starting at the 8th place.

An example of a non-repeating pattern could be

0 1 0 1 1 0 1 1 1 0 1 1 1 1 0 1 1 1 1 1 0 1 1 1 1 1... the string of 1's
getting longer and longer

Codes with repeating patterns will lead us to the number system known as the rational numbers, whereas the non-repeating ones will introduce the so-called irrational numbers.

There are many other ways of defining a property, for example we might require that a code of a position belonging to our set should not contain more than three 1's. Or we might require that after every run of two 1's there must be a 0.

It must be stressed that we are not yet dealing with numbers. Before we can talk about numbers we have to introduce the notion of measure, which will provide the answer to the question : "How much?", after which all the above can be applied to numbers.

Every position in the sequence given in section (7), starting with the second one, belongs to the interval

$$(1\,0\,0\,0\,0\ldots,1\,1\,0\,0\,0\,0\ldots)$$

The members of the sequence could also be defined as a set by the following property:

The code must begin with 1 0, after which any number of 1's are placed, followed by an indefinite succession of 0's. Of course this does not define the sequence as it does not tell you which member comes before which. A sequence has an order, a set does not.

NEIGHBOURHOODS, ISOLATION, AND POINTS OF ACCUMULATION

Given any position, there are many intervals to which it belongs. The set of all such intervals is called the neighbourhood of the position in question. So any interval in which a certain position is situated forms part of the neighbourhood of that position.

Let us say that we have defined a set of positions, let us call it S. Let us think of a position X, which may or may not be an element of S. Let us think of an interval that has X in it. Let us assume that there is at least one element of S in this interval. It may happen that however we choose our interval containing X, there will always be an element of S (other than X if X is an element of S) in it. In such a case we say that X is not isolated from S. If there is an interval containing X which does not contain any elements of S (except X in the case X is an element of S) then we say that X is isolated from S. Let us make up an example:

Consider the set whose codes contain just two 1's, the rest of the symbols being 0's.

This set will contain positions with codes such as

$$1\,0\,1\,0\,0\ldots,0\,1\,1\,0\,0\,0\ldots,0\,0\,1\,0\,0\,0\,0\,0\,1\,0\,0\,0\ldots$$

Whatever open interval we take in which 1 0 0 0 0 0 … is included, the interval will contain members of our set. So 1 0 0 0 0 0 0 … is not isolated from our set.

A position not isolated from a set seems to have a lot of members of the set densely congregating around that position. For this reason we call a non-isolated position a position (or point) of accumulation of the set. Such a point of accumulation may or may not be a member of our set.

If there is only one point of accumulation associated to a set, this point of accumulation is called the limit of the set.

If you want to convince yourself of such "dense accumulation", you can go through the following steps:

Suppose that every interval containing X also contains a member of the set S. Call such a position X_1., we could exclude X_1 yet include X by choosing another interval. There will be an X_2 of S in this new interval which will still include X. Then we choose another interval in the neighbourhood of X excluding X_2 which will have an X_3 of S. Clearly this can go on indefinitely and so we can find a great number of members of X in the neighbourhood of X (in fact an "infinite" number!).

Let me go back to my remarks about increasing sequences. By an increasing sequence it is meant that out of any successive members of a sequence the "next one" is always further from the start than the "previous one". Naturally, there are also decreasing sequences, whose members get "nearer and nearer" to the start 0 0 0 0 0 0 0......

A sequence which is ever increasing or ever decreasing but never stationary (no two successive members can be equal) is called strictly monotonic.

A strictly monotonic sequence must always proceed in the same sense, namely either always forward or always backward. It can never repeat (meaning of "strictly") and it can never turn round (meaning of "monotonic").

Let us recapitulate: we have introduced the idea of the neighbourhood of a position, which is the set of open intervals covering any particular position. "Covering" means that the position belongs to the interval. We have also defined sets of positions as all positions of a certain kind, of having some common property. Then we went on to define a relation of a position to a set of positions, namely on the one hand the position in question may be isolable from the set or it may be non-isolable from the set we are considering. A position which is non-isolable from a set is called a position of accumulation (or point of accumulation) of that set.

If just one such non-isolable of a set exists, the position is called the limit of the set.

THE BOLZANO–WEIERSTRASS THEOREM

Now it will be interesting to ask the question whether any infinite set of positions between start and finish has at least one point of accumulation.

Perhaps we can ask an easier question first:

Does a strictly monotonic sequence have a point of accumulation?

To answer this question, we can proceed in the following way:

Let us remember that each member of any proposed monotonic sequence is also a sequence of zeros and one's. Each member of our sequence, which we could consider monotonic increasing has a first symbol (either 0 or 1), a second symbol (either 0 or 1), a third symbol (either 0 or 1) and so on.

Let us look at the first symbol of the first position, then the first symbol of the second position, then the first symbol of the third position, and so on down the line, then only three possible cases can occur:

(i)   either all the symbols are zeros
(ii) or all the symbols are ones
(iii) or a lot of zeros followed by a lot of ones

That is because once you have a one, it is not possible to have a zero as in such a case the sequence would have a position nearer the start than the previous position!

So the first symbols "settle down" to a final value at some point, which might happen right at the beginning or at any time after (when you get the first 1 after a whole lot of 0's). Let us call this final value A_1. From here on look at the second symbols. From this point on these second symbols must in either all zeros or all ones or zeros followed by a lot of ones. So these second symbols will now also "settle down" to a final value which will not change. Let us call this final value A_2. After this look at the third symbols which will also "settle down" to a final value A_3. Continuing in this way we find a position with the code

$$A_1 \ A_2 \ A_3 \ A_4 \ A_5 \ \dots\dots\dots\dots\dots\dots = A$$

It can fairly easily seen that this position is not isolable from our sequence.

It is not hard to see that whatever interval we take that includes A, there will always be some members of our sequence in it, because by definition because this "final" limiting position is determined by the sequence itself and its members partially coincide with it more and more in the first, in the second, in the third, in the fourth place and so on as we proceed in the sequence of positions.

So in the case of a strictly increasing monotonic sequence there is always a position of accumulation.

A very similar argument will show us that in the case of a strictly decreasing monotonic sequence there will also be a position of accumulation.

If we succeed in showing that any infinite set all of whose members are between start and finish will contain either a strictly increasing or a strictly decreasing monotonic sequence, we shall have shown that every such infinite set will have a position of accumulation associated to it.

For grammatical simplicity, I shall use the words "greater" and "smaller" instead of "nearer to the finish" and "nearer to the start". "Greatest" will mean "there is no greater".

Here is the proof. Read it slowly and make sure you agree with all the points.

> Let S be an infinite set of positions, all positions being between start and finish.

Either the set does or it does not have a greatest member.

If it does not have a greatest member, then we can find an increasing sequence among its members, as whichever member we choose, there will be a greater one. Since there is no greatest, there will be a greater one than that one, and for the same reason there will be an even greater one than that one and so on, we can clearly construct an increasing sequence.

If it does have a greatest member, then consider the Set S_1, which is the set S with the greatest of S taken away. Again S_1 may or may not have a greatest. If it does not, we can construct an increasing sequence as before. If it does have a greatest, then consider the set S_2 which is S_1 deprived of its greatest member. Then S_2 may or may not have a greatest member. If it does not, we know we can construct an increasing sequence. If it does, take it away and thus form the set S_3 which again may or may not have a greatest member. We might go on with this procedure until we come to a set which does not have a greatest member. Of course this might never happen. So there are two possible outcomes:

(i) the process ends by reaching a set with no greatest member,
(ii) the process never ends

In case (i) we know we can construct an increasing sequence. In case (ii) the sequence of elements we keep taking away for forming new sets gives us a monotone decreasing sequence.

So in every possible eventuality, we can find a monotonic sequence in our infinite set S.

So every infinite set between start and finish contains a strictly monotonic sequence, and since to every strictly monotonic sequence we can associate a position of accumulation, it follows that we can associate a position of accumulation to every infinite set of positions contained between start and finish.

The above is known as the Bolzano–Weierstrass theorem.

Figure 4.4 is a diagrammatic representation of the argument showing that in every infinite set between start and finish there is always a strictly monotonic sequence.

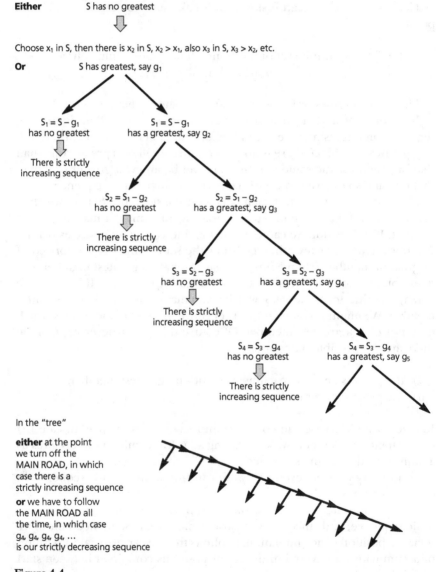

Either S has no greatest

Choose x_1 in S, then there is x_2 in S, $x_2 > x_1$, also x_3 in S, $x_3 > x_2$, etc.

Or S has greatest, say g_1

$S_1 = S - g_1$
has no greatest

There is strictly
increasing sequence

$S_1 = S - g_1$
has a greatest, say g_2

$S_2 = S_1 - g_2$
has no greatest

There is strictly
increasing sequence

$S_2 = S_1 - g_2$
has a greatest, say g_3

$S_3 = S_2 - g_3$
has no greatest

There is strictly
increasing sequence

$S_3 = S_2 - g_3$
has a greatest, say g_4

$S_4 = S_3 - g_4$
has no greatest

There is strictly
increasing sequence

$S_4 = S_3 - g_4$
has a greatest, say g_5

In the "tree"

either at the point
we turn off the
MAIN ROAD, in which
case there is a
strictly increasing sequence

or we have to follow
the MAIN ROAD all
the time, in which case
$g_1, g_2, g_3, g_4 \ldots$
is our strictly decreasing sequence

Figure 4.4

CHAPTER 5

GAMES LEADING TO THE TETRAHEDRON

Zoltan Paul Dienes

INTRODUCTION

You will need to have

a. a set of logic blocks (attribute blocks)
b. a small toy car
c. a four-colour board for the car-game (to be described later)
d. a regular tetrahedron made of cardboard, with coloured circles and squares on the faces,
e. twenty red, twenty blue, twenty yellow and twenty green counters.

In the sequel you will see instructions for making the board and the tetrahedron.

In all of the games there will be "moves", which you will have to carry out according to the rules of each game.

In some of the games players will have to reach a certain goal in the smallest possible number of moves. Other games will be games about games, in the sense that players will need to look for ways in which different games they have played are like each other. So these will be meta-games!

Mathematics Education and the Legacy of Zoltan Paul Dienes, pages 67–93

The study of the tetrahedron is based on the six stage theory of learning abstract ideas in mathematics, these being

1. an initial random interaction with a given environment;
2. the learning of "games" with rules which are so made up as to "copy" the abstract ideas to be learned;
3. the realization that all the "games" have the same construction, the same inner structure;
4. the representation of all the "games" with the same structure through a visual map;
5. the development of a language for describing the map, a process known as symbolization;
6. the tidying up of the description into initial descriptions (axioms) and consequences thereof (theorems), together with precise "rules to derive consequences" giving rise to chains linking the axioms to theorems (proofs); this process is known as formalization.

Each of the above stages depends on the completion of the previous stage and so by following these stages we are able to bring into being an organic growth pattern of learning.

GAME 1: WHO PUTS DOWN THE MOST BLOCKS? (STAGE 2)

Take twelve blocks from a set of logic blocks, not just at random but making sure that your twelve blocks are either the thin circles and the thin squares, or you could have the thick circles and the thick squares.

The game is played like this:

Choose two of the rules described on the next page.

Note the colour of each of your chosen rules.

Make sure you have some counters of the colours of the rules you have chosen.

Put down any one of the twelve blocks as the first block.

Choose one of your rules as the first one you will use.

Put that colour counter next to your first block.

Put down the second block, as your chosen rule tells you.

Choose a rule again. This can be the same rule as before or the other one.

Put that colour counter next to your second block.

Put down the third block using the rule chosen.

Go on putting down counters and blocks, each time using one or the other of your chosen rules, until you find no way of putting a block down, because the block you get by using your rules has already been put down!

Count how many counters you have put down. This is your score. Put the twelve blocks and counters back in a pile.

Now it is your friend's turn.

He must use the same two rules that you have used.

He must put down coloured counters and blocks, in the same way that you did, but try to put more blocks and counters down The player who puts the most counters down is the winner.

One player can challenge the other player if he or she thinks that the other player has used a rule that is not allowed or has used an allowed rule incorrectly. If the challenger is correct in his challenge, he or she is the winner and the game ends there.

The Four Rules of the Game

Here are the rules:

(i) *The Red Rule*
The next block must be:

a. the same colour as the previous block,
b. if the previous block is yellow, the next block must be different only in size from the previous one,
c. if the previous block is blue, the next block must be different in shape and in size from the previous one,
d. if the previous block is red, the next block must be different only in shape from the previous block.

(ii) *The Blue Rule*
The next block must different from the previous block in *colour only*.

a. If the previous block is yellow, the next block is blue,
b. if the previous block is blue, the next block red,
c. if the previous block is red, the next block is yellow.

(iii) *The Yellow Rule*

   a. The same colour change as in the blue rule, namely yellow is followed by blue, blue by red and red by yellow, but *also*:

   b. If the previous block is yellow, the next block is also different in
   shape and size from the previous block,

   c. if the previous block is blue, the next block is also different in
   shape from the previous block,

   d. if the previous block is red, the next block is also different in
   size from the previous block.

(iv) *The Green Rule*

Only the colour changes but yellow is followed by red, red by blue and blue by yellow.

It is possible to choose the rules in such a way that all the blocks are used. So the highest possible score for a turn is 11, as eleven counters would lead you to the twelfth or last block

Here is a chain made by a player. Find what rules he used each time, and whether they were correctly used!

small blue square	small red square	small yellow square	big yellow square	big blue square	small blue circle	small red circle

Why could he not go any further?

GAME 2: THE START TO FINISH GAME (STAGE 2)

Two rules are chosen in agreement with your friend .

   Now try to make a chain of blocks, using the two chosen rules, that
   will bring you from one of the blocks to the other block.

   Your score is the number of counters you have put down.

   Now it is your friend's turn.

   Your friend has to put down a chain of blocks and counters, using
   the same rules as yourself, but try to make the chain shorter.

   Your friend's score is the number of counters he has put down.

   The winner is the player with the lower score.

Here is an example of how a game might have been played:
   Players choose the big blue circle and the big blue square.

Player 1 makes the chain:

> Big blue circle → blue → Big red circle → blue → Big yellow circle →
> red → small yellow circle → blue → small blue circle → red → big
> blue square

and has placed five counters so his score is 5.

Player 2 on the other hand does this:

> Big blue circle → blue → Big red circle → red → Big red square →
> blue → big yellow square → blue → big blue square

This player has only used four counters, so his score is 4. So player 2 wins, as his is the lower score.

You should be smart enough to do some "hard" tasks. The "harder" tasks need longer chains. You are reminded that what might turn out to be a long chain with two given rules, might be a very short one with two other rules!

THE THREE MEALS GAME

To play this game you will need a rectangular table and four people, or if you prefer, a yellow, a blue and red rectangle from a set of logic blocks, as well as all the small thick blocks, or all the small thin blocks.

The yellow rectangle stands for the real table, but at breakfast time. The blue rectangle stands for the table at lunch time and the red rectangle for the table at supper time.

The small blocks stand for the four people. We can pretend that the yellow blocks are the four people at breakfast time, the blue blocks the four people at lunch time and the red blocks the four people at supper time. The "people" always sit along the two long sides of the "table", nobody ever sits at the two ends.

GAME 3: THE SEATING PROBLEM (STAGE 2)

Try to seat the "people" at the three meals in a way so that everybody will have sat next to a different person at each of the three meals, and also so that everybody will have sat opposite different people at each of the three meals.

For each meal, the first person can sit down anywhere, but you have to be careful how you seat the other three.

GAME 4: WHO SHOULD SERVE AT EACH MEAL?
(STAGE 2)

Choose one person who will be the one to serve at breakfast. But this person does not want to serve at all the other meals, so the four persons agree on a rule according to which if a person serves at one of the meals, then the person sitting next to him or her, will be serving at the next meal. They call this rule the "next rule"

Using this rule, it is soon found that whoever serves breakfast one day, will again be serving breakfast the next day, and whoever was serving lunch one day, will do so again the day, the same thing being true for supper. This means that one person gets away with not doing any work. The four people discuss this problem and decide to change the rule as follows: According to this rule, if a person serves at one of the meals, the person sitting opposite him or her will be serving at the next meal. They call this rule the "opposite rule".

Using this rule, it is true that different people will now be working, in fact the one that got away with it last time, does not do so this time, but it is found that another person gets away with doing no work if this same rule is used all the time. Try to find a way of using the rules, sometimes the "next rule" and sometimes the "opposite rule" so that the serving is done fairly. How will you arrange the use of the rules so that everybody gets to serve at each one of the meals just once!

Try to work out "fair" shifts for four days!

GAME 5: THE LONGEST SERVING CHAIN GAME
(STAGE 2)

Maybe you have not been able to solve the problem of arranging the serving quite fairly, in which case you can play a game like this:

You take it in turn to say ways of using the two rules so that it takes as long as possible before you serve again at a meal at which you have already served. The player who finds the longest "chain", is the winner.

Here is a way the game might have been played:

The seating arrangement for the three meals was like this:

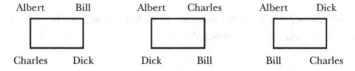

where everybody sits next to a different person at each meal and everybody sits opposite a different person at each meal.

Player 1 suggests this: (NX is next, OP is opposite)

Albert	Dick	Bill	Dick	Charles	Dick	Bill	Albert	Charles
NX	OP	NX	OP	NX	NX	OP	OP	
supper	breakfast	lunch	supper	breakfast	lunch	supper	breakfast	lunch

This is where it ends, because either rule would get a person to serve at a meal at which he has already served! If you are player 2, can you beat player 1?

Here is a way to beat player 1, but don't look at it, try to work it out for yourself!

[next- opp- opp- next- next- opp- opp- next- next- opp- opp]

If you have "peeped," try to find another way to get everyone to serve at every meal just once during a run of four days!

Invent some other rules for the service. What would happen if you had a maid to do the work at every other meal? Or if you have a headache, you could change service during a meal too!

THE CAR GAME
GAME 6: MAKING THE BOARD FOR THE CAR GAME
(STAGE 2)

The best way to make the board is to use four different kinds of *round* beer-mats (coasters). Maybe an adult friend you know can get you some from a bar! Or you could collect the metal tops of frozen juice containers. You could use unifix cubes to put on them, having chosen four colours.

When have your four different kinds of beer-mats, then put them together on a flat surface such as the floor or a large table, as closely packed together as they will go.

Make sure though that two mats of the same kind or juice covers with the same color cubes never touch each other, nor are they placed "opposite" each other.

You can also use square mats, although hexagonal mats (with six sides each) are probably better. Let us suppose that you have some circular mats, and that they are of four colours:

(1) yellow, (2) blue, (3) red, (4) green

You can put them out like this (Figure 5.1) to make a board: I have used clear for yellow, "squares" for red, level stripes ("lake") for blue, and "vertical" stripes (trees?) for green.

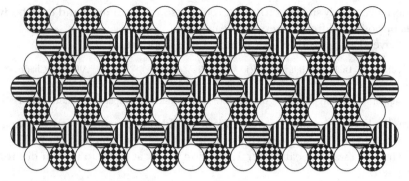

Figure 5.1

You will see on this board that mats next to each other as well as "oppo-site" mats have different colours.

Now take a little toy car. It must be a car whose front wheels can rest on one mat and whose rear wheels can rest on a mat next to it. "Next" does not only mean towards the right or towards the left on the board, the car can also be placed going "upwards to the right" or "upwards to the left" or "downwards to the right" or "downwards to the left".

You can explain how the car is placed by stating the colour of the mat where the rear wheels are, then the colour of the mat where the front wheels are. For example if you wanted your car to be "upwards right" it would have to go on blue–red, or on red–blue, or on green–yellow, or on yellow–green.

Find how many different colour positions the car can have.

GAME 7: HOW TO MOVE THE CAR (STAGE 2)

Here are some rules for moving the car about on the board. We can start with four different moves:

(i) The Neutral Move

You can move the car to another position in such a way that both the front and the rear wheels end up on the same colours as the colours on which they were before.

The neutral move does not count as a move when we count how many moves it takes to get from a position to another position.

(ii) The Forward Move

The car is moved forward without turning either to the right or to the left. The rear comes up to where the front was.

For example a car on blue–red comes to red–blue, or from red–green it comes to green–red. The front and rear "mats" simply change colours.

(iii) The Gentle Turn to the Right

The car is moved round the gentle bend turning it to the right, the rear again moving to the space where the front was before you moved it.

For example a car on blue–red would go to red–yellow. If you went on doing the gentle turn to the right, you would find that the car would circle round the green mat all the time, but would never go on any green mat.

(iv) The Gentle Turn to the Left

This is done in the same way as the previous move, except that the turn is to the left instead of to the right.

For example a car on blue–red would go on to red–green, then on to green–blue, if you did it again, and if you went on with it, it would circle the yellow mat, but never actually go on a yellow mat!

Practise the moves by doing some of the following:

Place the car on the board in any position you choose.

Give your friend one of the moves.

Your friend then must move the car, using the move you chose.

Then your friend gives you one of the four moves.

You now move the car, using the move your friend chose.

Then you can go on to saying two moves one after the other, and the other player has to carry out the moves, but in the order in which the moves were given. When you get good at this, try three or even four moves!

GAME 8: THE LONGEST RUN GAME (STAGE 2)

Choose two of the moves (ii), (iii), (iv) .

These will be the "allowed moves"

The neutral move is always allowed, and does not count as a move when you count the number of moves you made.

Player 1 takes the car on a run, using only the allowed moves

Player 2 checks how the run is made, by writing down the colour pair where the car is after each move.

The run comes to an end when one of these colour pairs is repeated.

A colour pair in one order is counted as different from the same two colours in the opposite order.

The score is the number of moves made to reach the repeated colour pair.

Now player 2 takes the car for a run and player 1 is the one to write down how the run is made.

The player who can take the car for the longest run wins.

The longest run takes you through all the possible positions.

Don't read the next example if you want to do it yourself!

Here is how one player got through all the positions by choosing the Forward and the Gentle turn to the right. Here is his game:

gy → FWD → yg → RT → gb → RT → by → FWD → yb → RT →
br → RT → ry → FWD → yr → RT → rg → FWD → gr → RT → rb →
RT → bg

In any case do check that the player who played this played it correctly.

Then try to take the car for a run using right and left turns only, never any forwards! Can you get all the positions in? Then try using the left turn and the forward turn. How long a run can you make with those two moves?

GAME 9: THE SHORTEST RUN GAME (STAGE 2)

Two of the three moves are picked as "allowed moves"

Player 1 picks two positions, a starting one and an end one, by saying the colour pair of each one.

Player 2 has to go from the starting position to the end position in the smallest number of allowable moves.

Player 1 now has to try to beat Player 2 by doing the run in a smaller number of moves.

The player who gets from one position to the other in the smaller number of moves is the winner.

Now player 2 picks the two positions and the game starts again, each player trying to join the two positions in the smallest number of allowed moves.

Later you can ask how you get back from the end position to the starting position. Will the same chain of moves do it, or do you need another chain? Or is it sometimes the same chain and other times a different one that takes you back to start?

For example players might want to run the car from

Blue–Red to Blue–Yellow
using Forward and Right

One player makes the run:

Forward–Right–Forward–Right

while the other makes the run:

Right–Right–Forward

Do they both arrive at Blue–Yellow? Check it carefully on your board with your car.

Do you ever need to use more than four moves to get from a position to another? Or are there some problems that take longer runs?

If you do Right then Left then Right then Left and so on, for how long do you have to go on before you get back to your starting colour pair?

How many times do you have to do Right then Forward before you get back to your starting colour pair?

If you want to make both your colours different, how many moves do you need? For example how do you get from

Red–Yellow to Blue–Green?

Ask yourself all sorts of questions like these and try to find the answers.

GAME 10: A GAME ABOUT THE GAMES (STAGE 3)

How are the games like each other? Is there a common thread running through them?

In the blocks games you were playing with 12 blocks, in the meals games you were solving problems in which there were 12 ways of serving meals, and in the car games you had to move the car between two positions, and there were 12 possible positions, shown as colour pairs. So all the games so far have been "12-games".

Try to make up a way of passing from something you do in one game to something you do in one of the other games, by saying, for example, which block "goes with" which person serving at which meal, or which block "goes with" which colour pair in the car game?

Then say which move in one game means doing which move in the other game.

If there are not enough moves in a game, you can always make up some more moves.

If two of you "play", then each of you should make up his or her own "dictionary" for "translating" each game into the other, and then compare your dictionaries with each other.

For example here is a dictionary made up by a player:

big red square = red–green	small red square = green–red
big red circle = yellow–blue	small red circle = blue–yellow
big blue square = yellow–red	small blue square = blue–green
big blue circle = red–yellow	small blue circle = green–blue
big yel. Square = green–yellow	small yel. Square = red–blue
big yel. Circle = blue–red	small yel. circle = yellow–green
Forward = Red rule	
Right turn = Blue rule	

We can test the dictionary by seeing if it "translates" what is correct in one game into what is correct in the other game. For example:

Big yellow   Big blue
→ blue →
circle   circle

translates into the "sentence":

(blue red) RIGHT (red yellow)

which you can check is correct by using your toy car.

Try some other sentences that are "true" in one of the games translate, and check that the translated sentence is true in the other game. Then make up your own dictionaries and work them!

THE TETRAHEDRON GAME
GAME 11: HOW TO MAKE A TETRAHEDRON? (STAGE 2)

Take a fairly big piece of cardboard and cut out an equilateral triangle from it. If you do not know how to draw one, just take one of the big triangles from a set of logic blocks and trace it on to the cardboard.

Find the point along each side of your triangle which is exactly half way between the corners (vertices). Mark them carefully.

Join these points to each other across your triangle.

You will now have four smaller triangles into which your big triangle has been divided.

Use the lines you drew across the big triangle as folds, folding the three triangles that end in the corners of the big triangle.

If you have done your drawing accurately, these smaller triangle should "stand up" at an angle and come together to make a kind of tent.

Use some sticky tape to make your "tent" solid.

Your "tent" is a model (Figure 5.2) of what is called a *tetrahedron*

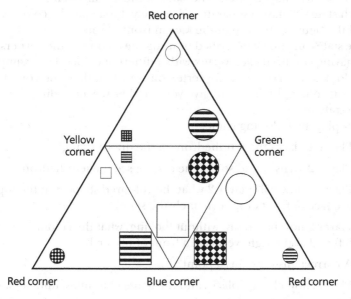

Figure 5.2

Colour the four corners of the "tent" as shown in the diagram, then also draw the coloured circles and squares exactly as shown above. Then fold the "tent" along the lines drawn inside the big triangle but in such a way as to leave the coloured figures on the outside. Now stick the "tent" together with adhesive tape and you will have a model of a tetrahedron, ready for playing the games to be described.

The four triangles making up the "tent" or tetrahedron are called the *faces*.

The lines at which two faces meet are called the *edges*.

The pointed corners are called the *vertices*.

GAME 12: HOW TO USE THE DRAWINGS ON YOUR TETRAHEDRON (STAGE 2)

Put your tetrahedron down on a piece of paper and draw round the triangle that it is standing on.

Always put the tetrahedron back so as one of the faces fits exactly onto this triangle you have drawn.

The coloured shape you see at the top of the face in front of you can be called the *state* of the tetrahedron.

There are twelve ways of placing the tetrahedron in front of you so that it stands exactly in the triangle you have drawn.

Each of these ways is one of the *states* of the tetrahedron.

Each state will have a separate *code* namely the shape, the colour and the size of the figure at the top of the face in front of you.

The state can also be described by using the colours at the corners (vertices), having coloured each vertex with a different colour. For example you could look at the colour at the vertex on your left, then the colour of the vertex on your right. This will give you another way of coding the state of your tetrahedron.

Now play the following game:

Player 1 places the tetrahedron on the drawn triangle.

Player 2 turns round so that he cannot see the tetrahedron.

Player 1 now tells player 2 what the coloured shape is at the top of the face in front of him (namely the state)

Player 2 now has to say, without looking, what the colours are at the left and at the right vertices in front of player 1.

A correct answer gains a point.

Players now change places and the game continues, until a player has scored three points.

The game can also be played by a player giving the left and the right colours, the other player having to say what the coloured shape is at the top of the face in front of the first player. Here we are encroaching on Stage 3 by comparing the two ways of representing states.

If you find difficulties with "guessing", go back to your dictionary between the blocks game and the car game. You might get a big surprise!

GAME 13: HOW TO MOVE THE TETRAHEDRON (STAGE 2)

Here are some suggested rules for turning the tetrahedron around. After each turn the tetrahedron must be replaced in the space in which it was before it was turned.

a. Hold the tetrahedron with your right hand at the upper vertex and turn it slowly as though you were turning right handed screw. Stop when you see another face in front of you.
b. Hold the tetrahedron at the middle of the level edge in front of you with the fingers of one hand and at the middle of the edge sloping away from you with the fingers of the other hand and

TWIDDLE IT ONLY WITH YOUR FINGERS
WITHOUT MOVING YOUR HANDS OR YOUR ARMS

turning the tetrahedron slowly, until you see another face in front of you.
c. Hold the top vertex of the tetrahedron with you left hand and turn it as though you were turning a left handed screw. Stop when you see another face in front of you.
d. Pick up the tetrahedron and turn the face in front of you in the clockwise sense. Stop when you see another shape at the top of a face.

You could also hold the tetrahedron with your left hand at the vertex furthest away from you and turn it as though it were a left handed screw. It will have the same effect!

For each of these rules the coloured shape you see in front of you will change in certain definite ways, obeying certain rules.

Try to find these rules. Are any of them the same rules that you used in the blocks game?

Find more ways of turning the tetrahedron (there are eleven altogether, not counting the NO MOVE), and find in what ways the colours, shapes and sizes change for each new turn that you find.

Now, instead of looking at the coloured shapes, look at the left and right vertices.

Find a rule in the car game for each turn you have found for your tetrahedron. The car rule should change the colour pairs in the car game in just the same way as the turning rule changes the left and right colours on your tetrahedron.

If you do not have enough car rules, invent some more. For example the car could skid to the right or to the left, with all the wheels or just with the front wheels or just with the rear. wheels.

GAME 14: GOING THROUGH ALL THE POSITIONS
(STAGE 2)

See what you can do with the moves (a) and (b).

Try to go through all the states of the tetrahedron without repeating any, using only moves (a) and (b).

Here is a solution, but don't look at it if you want to work it out yourself. If you cannot or do not feel like it, then just check that the chain of moves given here does take you through all the 12 states of the tetrahedron, without repeating any.

[(b) (a) (a) (b) (a) (a) (b) (a) (b) (a) (a)]

Find some other chains of moves using only (a) and (b), which will take you through all the 12 states.

Now choose two other moves among the four moves

(a), (b), (c), (d)

and again try to find a chain that will take you through all the states. Is it always possible?

Some of the moves get you back to the starting state when you have used them just twice.

These are *half turns* of the tetrahedron.

Other moves need to be done three times over, before they get you back to the starting state.

These are called *thirds of a turn* of the tetrahedron.

Make sure you know which of the moves are half turns and which are thirds of a turn.

If you go on doing the move (a), then the move (b), then the move (a), then the move (b) and so on, how long will it be before you come back to your starting state?

Decide on a starting state.

Do the moves (b), (a), (b) one after the other.

Note the state you get to.

Now use the long chain that takes you through all the states, starting again from the original starting state.

Do you get to the same state as when you did (b)(a)(b)?

GAME 15: THE START TO FINISH PROBLEM
(STAGE 2)

Choose any two states by choosing two coloured shapes or two colour pairs. One will be your starting state, the other your end state.

Choose two moves that you allow yourself to use.

Try to get from your starting state to your end state in the smallest possible number of moves, using the moves you have allowed yourself to use.

The player who manages in the smallest number of moves is the winner.

You might wonder what happens if you choose more than two allowable moves. Naturally, in some cases, you will then be able to reach your "goal" in a smaller number of moves than you were able to do with only two allowable moves.

If you have worked out all the eleven moves, then it should be possible to go from any state of the tetrahedron to any other state of the tetrahedron in just one move.

A game can be played in which two states are chosen by one player and the other player must say which of the eleven moves will take the tetrahedron from one of these to the other. You might even insist that you should be able to say which move will take the tetrahedron *back* from the second state to the first one.

Two such moves are called *inverse moves.*

Another way of playing with the eleven moves is to give a state and a move, and players must say which state would be reached if the tetrahedron were moved from the given state, using the given move.

Here are the eleven moves that you can use in the above games:

a. Turn the top vertex with your right hand,
b. Twiddle by the level front and sloping back
c. Turn the top vertex with your left hand,
d. Turn the back vertex with your left hand
e. Turn the back vertex with your right hand
f. Turn the left vertex with your left hand
g. Turn the right vertex with your right hand
h. Turn the left vertex with your right hand,
i. Turn the right vertex with your left hand
j. Twiddle the left level and sloping right edges
k. Twiddle the right level and sloping left edges

MAPPING THE TWELVE POSITIONS
GAME 16: USING A HALF TURN AND A THIRD OF A TURN
(STAGE 4)

Let us use the moves (a) and (b). If we wanted to know how to get from state to state in the shortest way, it would be good to have a picture or a map of all the twelve states, where the states would be joined by arrows, each type of arrow meaning one of the allowable moves. Two different kinds of arrow should be used, so we can tell one kind of move from the other kind

Figure 5.3 is such a map.

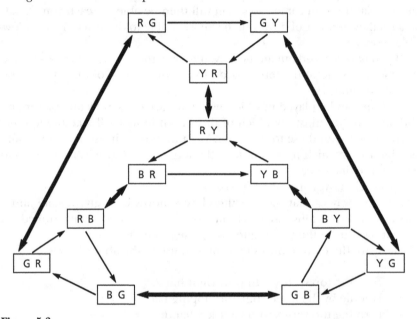

Figure 5.3

The states filled in are those of the car game, or if you like the states of the tetrahedron game, using the colour pairs.

The arrows pointing one way mean the move (a)

The arrows pointing both ways mean the move (b).

Make a map like this on a large sheet of paper or cardboard.

Fill in the spaces with the states of the blocks game, where the states are shown with the coloured shapes.

If you wanted to fill the map with the states of the meals game, you would have to have one of the moves as the one where you get a headache and change service with the person opposite, say, at the same meal. The other move could be the "next" or the "opposite" or even the "diagonal", at the next meal.

GAME 17: A MAP WHICH USES ONLY THIRDS OF A TURN (STAGE 4)

Here is another map (Figure 5.4):

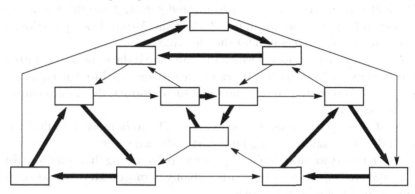

Figure 5.4

You could fill in the spaces with the twelve states of the tetrahedron. For example one of the arrows could be "Turn at the top with the right hand", while the other could be "Turn at the back with the right hand"

You can also fill this map in with the states of the meal game moves. It makes no difference which arrow is which move, as long as you always use the same arrow for the same move Imagine that the spaces are bus stops and that each "triangle" is a bus route. The buses doing the service on the thick arrows are double decker buses, those doing the service on the thin arrows are single deckers.

Choose two bus stops and try to get from one to the other, going on the smallest number of different buses.

Are there any journeys for which you need more than four different bus rides?

Find also different ways of getting from a given starting bus stop to a given end bus stop. For example if you and your friend start at the same stop and you take a double decker for two stops, then a single decker for two stops and your friend takes a single, then a double, then a single, then a double you will both end up at the same stop, but you only had to change buses once, but your friend had to change three times.

Two journeys that both begin and end at the same state are called *equivalent.*

THE DESCRIPTION OF THE MAPS (STAGE 5)
GAME 18: USING A HALF TURN AND A THIRD OF A TURN
(STAGE 5)

The half turn move corresponds to the red rule and the third of a turn corresponds to the blue rule in the blocks game. We could simple write the letter r for the red rule and b for the blue rule

A whole chain of moves in which the end state is the same as the starting state is a kind of "round trip". We can write the number 1 for such a "round trip chain". You should know by now that such a move is the "neutral move" or the "no move".

We shall write the equal sign between two chains of moves if both chains reach the same state assuming both start at the same state.

Now let us try to "describe" one of our maps by writing that certain chains of moves are "equal" to certain other chains of moves. Here are some examples "describing" our first map:

$$r\,r = 1, b\,b\,b = 1, b\,r\,b\,r\,b\,r = 1, r\,b\,r\,b\,r\,b = 1$$
$$r\,b\,r = b\,b\,r\,b\,b, b\,r\,b = r\,b\,b\,r,\ \text{and so on.}$$

What we are saying, in a new kind of language, is that

  (i) "If you take the arrow pointing both ways twice, we shall get back to where we started"
 (ii) "If we travel along the arrow pointing one way, we shall be back where we started after using such an arrow three times."
(iii) "If we use a one way arrow, then a two way arrow, then again a one way arrow, then a two way arrow, then again a one way arrow and then a two way arrow, we shall get back to where we started."
(iv) "If we take a double pointing arrow, then a single pointing arrow and then again a double pointing arrow, we shall get to the same state as if we had taken two single pointed arrows, one double pointing arrow and then again two single pointing arrows."

I am sure you will agree that writing

$$r\,b\,r = b\,b\,r\,b\,b$$

is a lot simpler than writing it out fully!

Try to find such "equivalent journeys" on both your maps and write them down in your notebook. The more you find, the better you will have described your map or maps.

GAME 19: USING TWO DIFFERENT THIRDS OF A TURN (STAGE 5)

Why not use the moves (a) and (e) for turning the tetrahedron?

Then we might as well use the letters a and e meaning the two moves (a) and (e). Remember that

(a) means: "Turn it with your right hand, holding it at the top"

(e) means: "Turn it with right hand, holding it at the back"

We can "describe" our second map by writing some such things as the following:

$$a\,a\,a = 1, e\,e\,e = 1, a\,e\,a\,e\,a\,e = 1, e\,a\,e\,a\,e\,a = 1$$
$$e\,a\,a\,e\,e = a\,e, a\,e\,e\,a\,a = e\,a, \text{and so on}$$

Make sure you understand what these "equivalences" mean.

Take the tetrahedron and look at its states through the left and right vertices in front of you. You can say what state you are in by saying these colours, first the left colour, then the right colour.

Try out all these "equivalences" by turning and twiddling your tetrahedron.

For example a a a = 1 could mean that if you put tetrahedron down with the green vertex at the top,

$$Y\,B \rightarrow a \rightarrow B\,R \rightarrow a \rightarrow R\,Y \rightarrow a \rightarrow Y\,B$$

in other words, using move (a) three times over takes you back to your starting state.

The "equivalence" e a a e e = a e could mean that, again starting with the same state as before, the chain

$$Y\,B \rightarrow e \rightarrow G\,Y \rightarrow a \rightarrow Y\,R \rightarrow a \rightarrow R\,G \rightarrow e \rightarrow B\,R \rightarrow e \rightarrow G\,B$$

gets you to the same state as the chain

$$Y\,B \rightarrow a \rightarrow B\,R \rightarrow e \rightarrow G\,B$$

It seems that you can exchange the moves (a) and (e) and if you start with something true, you end up with something true. Of course, if you start with something untrue, you will end up with something also untrue by exchanging (a) and (e). Try it with some of the "equivalences" that you have found and check on the map. It goes without saying that you cannot exchange half turns with thirds of a turn and get away with it!

So the second map is more *symmetrical* than the first. In a way, each arrow is "mirrored" into the other arrow in the second map. This is not so in the case of the first map.

GAME 20:
PROBLEM: HOW TO DESCRIBE THE FIRST MAP FULLY!
(STAGE 6)

Let us just use the following three *equivalences:*

(i) r r = 1, (ii) b b b = 1, (iii) b r b r b r = 1

Let us find some way in which we can get to the other *equivalences* starting from the three we have chosen. For example, how could we get to

r b r = b b r b b

without the use of the map?

We could allow these rules for turning a chain of moves into another chain of moves, so that the new chain changes our state always in the same way as our first chain:

Rule 1. We can insert move 1 anywhere in a chain or remove move 1 from a chain. Remember that move 1 is a "round trip," getting you back from where you started.

We are allowed three such "round trips". listed under

(i), (ii), and (iii).

Rule 2. We can place an equal sign between any two chains, which are obtained from each other by using Rule 1

Let us see whether we can get to the chain

b b r b b
from the chain   r b r
using Rule 1 several times over.
Start with   b b r b b
By rule 1   b b r b b b b r b r b r
We know that b b b is a "round trip" so we can write
b b r r b r b r
but r r is also a "round trip" so we can write
b b b r b r

again b b b being a "round trip" we are left with

r b r

so by rule 2 we can write r b r = b b r b b

We have "proved" a new equivalence, using the two rules.

The first three equivalences are called the *axioms* the new equivalence we managed to get to is called a *theorem* and the way of getting there is called a *proof*.

Now try your hand at "proving" the equivalences A, B, C and D using only

(i) r r = 1, (ii) b b b = 1, (iii) b r b r b r = 1

(A) r = b r b r b, (B) b b = r b r b r

(C) r b b r b b = b r, (D) b b r b r b b r = r b b

Instead of making chains of moves, you could get the children to make rows of boys and girls from the class. The move r will mean a girl, and the move b will mean a boy.

So the rules for changing a row of boys and girls are these:

(i) Two girls can always be inserted anywhere in the row, or removed from anywhere in the row, as long as they are next to each other.
(ii) Three boys can always be inserted anywhere in the row, or removed from anywhere in the row, as long as all three are next to each other.
(iii) A row of a boy, of a girl, of a boy, of a girl, of a boy and of a girl can be inserted anywhere in the row, or removed from anywhere in the row.

For example we can change the row

by placing three boys at the start of the row, and we have

but the children in the "box" can be sent away, so we are left with just two boys.

"Prove" some of your equivalences using boys and girls!

GAME 21:
PROBLEM: HOW TO DESCRIBE THE SECOND MAP FULLY
(STAGE 6)

We can describe the second map using the axioms

(i) a a a = 1, (ii) e e e = 1, (iii) a e e a e e = 1

Our problem is whether we can "get to" every other equivalence that we can "see" on the map by either inserting or leaving out

a a a   or   e e e   or   a e e a e e

For example we can easily see that

a e a e a e = 1

is true on the map, but can this equivalence be "got at" by playing the inserting and leaving out game? Let us try and get rid of the whole chain by using the "round trips"

a a a,   e e e,   and   a e e a e e

START: a e a e a e, insert a e e a e e to get

a e a [a e e a e e] e a e, then remove e e e to get

a e a a e e a a e, then insert another a e e a e e to get

a e a a [a e e a e e] e e a a e, then remove a a a and e e e

a e e e a e a a e, then remove e e e and get

a a e a a e, then insert another a e e a e e and get

a a [a e e a e e] e a a e, then remove a a a and e e e to get

e e a a a e, then remove a a a to get

e e e, then remove e e e and finally we have

1 FINISH.

By Rule 2 we can write a e a e a e = 1

This last equivalence is a theorem that we have proved from the axioms

$$a\ a\ a = 1, e\ e\ e = 1, \text{and } a\ e\ e\ a\ e\ e = 1$$

It is interesting to note that if we choose the equivalences

$$a\ a\ a = 1, \quad e\ e\ e = 1 \quad \text{and} \quad a\ e\ a\ e\ a\ e = 1 \text{ as our AXIOMS,}$$

it will be *impossible* to *prove* that a e e a e e = 1!

Have you any idea why such a *proof* should not be possible?

GAME 22: JUST SOMETHING ELSE TO THINK ABOUT (STAGE 6 PLUS)

You have three meals a day, breakfast, lunch and supper and you can invite your friend on the day, a day later or two days later, since you have a three day shift of menus. So there are 9 meals

Day 1	Day 2	Day 3
breakfast	breakfast	breakfast
lunch	lunch	lunch
supper	supper	supper

Let us have the following "moves" in a game in which you keep inviting your friend to share a meal with you:

(a) come to the next meal, (aa) come to the meal after next,

(1) come to this meal I am having now or three days later for

the same meal, it will be the same menu!

(e) come again tomorrow for the same meal,

(ee) come again the day after tomorrow for the same meal,

(ae) come exactly a day after the next meal,

(aee) come exactly two days after the next meal,

(aae) come exactly a day after the meal after the next,

(aaee) come exactly two days after the meal after the next.

You could draw a map (Figure 5.5) of this game like this:

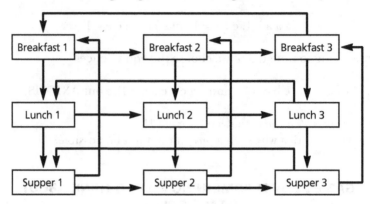

Figure 5.5

The "left-right" arrow means "a day later". We could use the letter a for this "move".

The "up-down" arrow means "the next meal". We could use the letter e for this "move"

It is true that a a a = 1, e e e = 1, a e a e a e = 1

but is it possible to "prove" that a e e a e e = 1?

from a a a = 1 and e e e = 1 and a e a e a e = 1?

This is hardly likely, since a e e a e e = 1 is false in this game! We have found a game in which the three axioms are true but a e e a e e = 1 is not, so the latter cannot be provable from the axioms!

It seems that our axioms

a a a = 1 and e e e = 1 and a e a e a e = 1

are *not sufficiently strong* for *proving*

a e e a e e = 1

and yet we found that the axioms

a a a = 1 and e e e = 1 and a e e a e e = 1

are *sufficiently strong* for *proving*

a e a e a e = 1

The "consequences" of the first set clearly do not cover all the equivalences that we can read off from the map of the tetrahedron. This set of axioms is *incomplete* as far as describing this map is concerned. One example has been enough to show *incompleteness*. It is much harder to prove *completeness*.

Is the second set of axioms enough to describe all the equivalences we can "see" in our tetrahedron map? In fact this is so, but to prove it goes beyond the scope of these notes! So I leave the reader pondering on how on earth such a thing could possibly be proved!

There are, in fact, many interesting problems to ponder over!

(i)   In how many other ways can we make up a 12-game? Will any such 12-game be "translatable" into one of our 12-games through a "dictionary" or not? If there are some "other" 12-games, how many are there?

(ii) What would a game be like that would handle the rotations and the symmetries of the regular hexagon?

(iii) What about a game that would "embody" all the rotations of the cube? What would happen if we also wanted to include its reflections? How many "states" would we get? How many "moves"?

(iv) Would the "cube-game" include in itself the "tetrahedron - game" as a "sub-game"? In other words, can the cube game be thought of as an extension of the tetrahedron game?

(v)   Can any such games be "played with" just using numbers and operations such as adding subtracting, multiplying and dividing? Which twelve numbers would you pick for a 12-game?

If you are troubled about the above or about anything else mathematical, e-mail me at

zoltan@zoltandienes.com

and I will do my best to help!

CHAPTER 6

COGNITIVE PSYCHOLOGY AND MATHEMATICS EDUCATION

Reflections on the Past and the Future

Lyn D. English
Queensland University of Technology, Australia

It has been well over a decade since I wrote the book, *Mathematics education: Models and processes* (1995), along with my coauthor, Graeme S. Halford. A good deal of what we wrote is still relevant to mathematics education today, as I indicate in this preface. But there have been many significant developments in the intervening years that have impacted on our discipline and indicate future directions for our field. I address some of these developments here.

ISSUES OF CONTINUED SIGNIFICANCE

Proponents of the period of Meaningful Learning (19030s and 1940s) advocated the development of mathematical learning with understanding,

Mathematics Education and the Legacy of Zoltan Paul Dienes, pages 95–105
Copyright © 2008 by Information Age Publishing
All rights of reproduction in any form reserved.

with William Brownell (e.g., 1945) emphasizing the importance of students appreciating and understanding the structure of mathematics. His recommendations are still highly relevant today: mathematics is the study of structure (Lesh & English, 2005). As highlighted in the National Council of Teachers of Mathematics' *Principles and Standards for School Mathematics* (NCTM 2000), students need to learn mathematics with understanding by actively building new knowledge from existing knowledge and experience. The curriculum must be "more than a collection of activities: it must be coherent, focused on important mathematics, and well articulated across the grades" (p. 14).

Van Engen's (e.g., 1949) advice, as well as that of the Gestaltists' (e.g., Wertheimer, 1959), is also highly pertinent to mathematics education today. For example, Van Engen emphasized the importance of developing students' ability to detect patterns in similar and seemingly diverse situations. In problem solving, students should identify the structure of a problem before searching for an answer. Likewise, the Gestaltists advocated the importance of productive thinking, as opposed to reproductive thinking, in the mathematics classroom (Wertheimer, 1959). A productive thinker grasps the structural relations in a problem or situation and then combines these parts into a dynamic whole. Such productive thinking can be encouraged by not giving students ready-made steps to solve given problems.

Developments during the period of the "New Mathematics" (1960s) still have significant input for mathematics education, despite the backlash they received in subsequent years. For example, Jerome Bruner's (1960) recommendation for students to progress through three levels of representation, namely, the enactive, iconic, and symbolic, is still sound advice for effective curriculum development today. Likewise, the ideas of Zoltan Dienes (e.g., 1960), whose recent interview appeared in the journal, *Mathematical Thinking and Learning* (volume 9, issue 1), remain highly applicable. For example, Dienes placed a strong focus on multiple embodiments and on cyclic patterns of learning where students progress from concrete to symbolic formats in developing an understanding of mathematical structures.

DEVELOPMENTS IN THE INTERVENING YEARS, 1995–2007

There have been numerous developments in the intervening years that have either changed the landscape of mathematics education and/or provide significant pointers for future growth of our discipline. These developments include, among others: (a) a decreased emphasis on constructivism as the dominant paradigm for the teaching and learning of mathematics (e.g., Goldin, in press; Lesh & Doerr, 2003); (b) new developments in the learning sciences, in particular, a focus on complexity theory (e.g., English, in press; Jacobson & Wilensky, 2006; Lesh, 2006); (c) an increased focus on mathematical reasoning and interdisciplinary modeling (e.g., English, 2006; Lesh & English, 2005; Lesh & Zawojewski, 2007); (d) a significant increase in research on the mathematics needed in various work place settings and the implications for mathematics education (e.g., Gainsburg, 2006; Hoyles, Noss, & Pozzi, 2001; Hoyles, Bakker, Kent, & Noss, in press); (e) a broadening of theoretical perspectives, including an increased focus on social-cultural-political aspects of mathematics education (e.g., Greer, Verschaffel, & Mukhopadhyay; in press; Gutstein, 2007); (g) developments in research methodology, in particular, a focus on design research (e.g., Cobb, Confrey, diSessa, Lehrer, & Schauble, 2003; Lesh & Sriraman, 2005; Shavelson, Phillips, Towne, & Feuer, 2003); and (h) the increased sophistication and availability of technology (e.g., Kaput, Hegedus & Lesh, in press; O'Neil & Perez, 2006).

DECREASED EMPHASIS ON CONSTRUCTIVISM

The dominant influence of constructivist theories on mathematics education has waned in recent years. In their edited volume, *Beyond constructivism: Models and modeling perspectives on mathematics problem solving, learning, and teaching*, Lesh and Doerr (2003) presented powerful arguments for why we need to move our field beyond the clutches of constructivist ideologies. For example, all of the goals of mathematics education do not need to be achieved through the processes of personal construction and not all the mathematics students learn need to be invented independently by students. In essence, construction is only one of many processes that contribute to the development of constructs.

Goldin (e.g., 2002, 2003, in press) has further highlighted the shortcomings of constructivist theories, explaining how radical constructivists:

> rejected on *a priori* grounds all that is external to the "worlds of experience" of human individuals. Excluding the very possibility of knowledge about the real world, they dismissed unknowable, "objective reality" to focus instead on "experiential reality." Mathematical structures, as abstractions apart from individual knowers and problem solvers, were likewise to be rejected. In advocating the (wholly subjective) idea of "viability" they dismissed its counterpart, the notion of (objective) validity (Goldin, in press).

Paradigms, such as constructivism, which became fashionable in mathematics education over recent decades, tended to dismiss or deny the integrity of fundamental aspects of mathematical and scientific knowledge. I agree with Goldin that "It is time to abandon, knowledgeably and thoughtfully, the dismissive fads and fashions – the 'isms' – in favor of a unifying, non-ideological, scientific and eclectic approach to research, an approach that allows for the *consilience* of knowledge across the disciplines" (2003, p. 176).

A models and modeling perspective provides one such unifying approach to research in mathematics education. I address this shortly but first wish to mention briefly some significant developments in complexity theory that have important, as yet untouched, implications for mathematics education.

DEVELOPMENTS IN THE LEARNING SCIENCES: A FOCUS ON COMPLEXITY

In the past decade or so, the learning sciences have seen substantial growth in research on complex systems and complexity theories (e.g., Jacobson & Wilensky, 2006; Lesh, 2006). Such research is yet to have an impact on mathematics education. Clearly, we cannot ignore this body of research if we are to prepare our students effectively for their future lives. Our students live in a world that is increasingly governed by complex systems that are dynamic, self-organizing, and continually adapting. Financial corporations, political parties, education systems, and the World Wide Web are just a few examples of complex systems. In the 21st century, such systems are becoming increasingly important in the everyday lives of both children and adults. For all citizens, an appreciation and understanding of the world as interlocked complex systems is critical for making effective decisions about one's life as both an individual and as a community member (Bar-Yam, 2004; Davis & Sumara, 2006; Jacobson & Wilensky, 2006; Lesh, 2006).

In basic terms, complexity is the study of systems of interconnected components whose behavior cannot be explained solely by the properties of their parts but from the behavior that arises from their interconnectedness; the field has led to significant scientific methodological advances (Sabelli, 2006). Educational leaders from different walks of life are emphasizing the need to develop students' abilities to deal with complex systems for success beyond school. These abilities include: constructing, describing, explaining, manipulating, and predicting complex systems (such as sophisticated buying, leasing, and loan plans); working on multi-phase and multi-component projects in which planning, monitoring, and communicating are critical for success; and adapting rapidly to ever-evolving conceptual tools (or complex artifacts) and resources (English, 2002; Gainsburg, 2006; Lesh & Doerr, 2003). One approach to developing such abilities is through mathematical modelling, which is central to the study of complexity and to modern science.

INCREASED FOCUS ON MATHEMATICAL REASONING
AND INTERDISCIPLINARY MODELING

Modelling is increasingly recognized as a powerful vehicle for not only promoting students' understanding of a wide range of key mathematical and scientific constructs, but also for helping them appreciate the potential of mathematics as a critical tool for analyzing important issues in their lives, communities, and society in general (Greer, Verschaffel, & Mukhopadhyay, in press; Romberg et al., 2005).

The terms, *models* and *modeling*, have been used variously in the literature, including with reference to solving word problems, conducting mathematical simulations, creating representations of problem situations, and creating internal, psychological representations while solving a particular problem (e.g., Doerr & Tripp, 1999; Gravemeijer, 1999; Greer, 1997; Lesh & Doerr, 2003; Romberg et al., 2005). In my research in recent years I have defined models as "systems of elements, operations, relationships, and rules that can be used to describe, explain, or predict the behavior of some other familiar system" (Doerr & English, 2003, p.112). From this perspective, modeling problems are realistically complex situations where the problem solver engages in mathematical thinking beyond the usual school experience and where the products to be generated often include complex artifacts or conceptual tools that are needed for some purpose, or to accomplish some goal (Lesh & Zawojewski, 2007).

Students' development of powerful models should be regarded as among the most significant goals of mathematics education (Lesh & Sriraman, 2005). Importantly, modeling needs to be integrated within the *elementary* school curriculum and not reserved for the secondary school years and beyond as it has been traditionally. My recent research has shown that elementary school children are indeed capable of developing their own models and sense-making systems for dealing with complex problem situations (e.g., English, 2006; English & Watters, 2005). Mathematics education needs to give greater attention to developing the mathematical modeling abilities of younger children, especially given the increasing importance of modeling beyond the classroom.

RESEARCH ON THE MATHEMATICS NEEDED
IN WORK PLACE SETTINGS: IMPLICATIONS
FOR MATHEMATICS EDUCATION

Numerous researchers and employer groups have expressed concerns that educators are not giving adequate attention to the understandings and abilities that are needed for success beyond school. Research suggests that although professionals in mathematics-related fields draw upon their school learning, they do so in a flexible and creative manner, unlike the way in which they experienced mathematics in their school days (Gainsburg, 2006; Hall, 1999; Hamilton, in press; Noss, Hoyles, & Pozzi, 2002; Zawojewski & McCarthy, 2007). Furthermore, this research has indicated that such professionals draw upon interdisciplinary knowledge in solving problems and communicating their findings.

The advent of digital technologies is also changing the nature of the mathematics needed in the work place, as I indicate later. These technological developments have led to both the addition of new mathematical competencies and the elimination of existing mathematical skills that were once part of the worker's toolkit (e.g., Jenkins, Clinton, Purushotma, & Weigel, 2006; Lombardi & Lombardi, 2007). Studies of the nature and role of mathematics used in the workplace and other everyday settings (e.g., nursing, engineering, grocery shopping, dieting, architecture, fish hatcheries) provide significant pointers for the future-orienting of mathematics education. In the final chapter of the second edition of the *Handbook of International Research in Mathematics Education* (English, in press), I list a number of powerful mathematical ideas for the 21st century that are indicated by these studies of mathematics in work place settings. Included in this list are: (a) working collaboratively on complex problems where planning, monitoring, and communicating are critical for success; (b) applying numerical and algebraic reasoning in an efficient, flexible, and creative manner; (c) generating, analyzing, operating on, and transforming complex data sets; (d) applying an understanding of core ideas from ratio and proportion, probability, rate, change, accumulation, continuity, and limit; (e) constructing, describing, explaining, manipulating, and predicting complex systems; and (f) thinking critically and being able to make sound judgments.

A BROADENING OF THEORETICAL PERSPECTIVES: AN INCREASED FOCUS ON SOCIAL-CULTURAL-POLITICAL ASPECTS OF MATHEMATICS EDUCATION

In recent years we have seen a major shift within the field of mathematics education from a mainly psychological and pedagogical perspective towards one that encompasses the historical, cultural, social, and political contexts of both mathematics and mathematics education (e.g., English, in press; Greer, Verschaffel, & Mukhopadhyay; in press). This multitude of factors is having an unprecedented impact on mathematics education and its research endeavours. Many of our current educational problems continue to be fuelled by opposing values held by policy makers, program developers, professional groups, and community organizations (Greer et al., in press; Skovsmose & Valero, in press). When mathematics is intertwined with human contexts and practices, it follows that social accountability must be applied to the discipline (D'Ambrosio, 2007; Greer & Mukhopadhyay, 2003; Mukhopadhyay & Greer, 2007; Gutstein, 2007). One potentially rich opportunity to address this issue lies in the increased emphasis on the inclusion of real-world problems in school curricula that involve data handling, statistical reasoning, and mathematical modeling and applications. There are numerous real-world examples where students can use mathematics to analyze socially and culturally relevant problems (Greer & Mukhopadhyay (2003). For example, Mukhopadhyay and Greer (2007) have outlined how the issue of gun violence, in particular as it impacts on students, can be analyzed in relation to its socio-political contexts using mathematics as a critical tool.

In the final section of this preface I review briefly design research, which in recent years has had a powerful impact on studies aimed at improving the teaching and learning of mathematics.

DEVELOPMENTS IN RESEARCH METHODOLOGY: A FOCUS ON DESIGN RESEARCH

In recent years, the field of mathematics education research has been viewed by several scholars as a design science akin to engineering and other emerging interdisciplinary fields (e.g., Cobb et al., 2003; Lesh & Sriraman, 2005; Hjalmarson & Lesh, in press). Design research typically involves creating opportunities for both "engineering" particular forms of learning and teaching and studying these forms systematically within the supportive contexts created (Cobb et al., 2003; Lesh & Clarke, 2000; Schorr & Koellner-Clarke, 2003). Such a process usually involves a series of "iterative design cycles," in which trial outcomes are iteratively tested and revised in progressing towards the improvement of mathematics teaching and learning (Lesh & Sriraman, 2005; Shavelson et al., 2003). Such developmental cycles leave auditable trails of documentation that reveal significant information about how and why the desired outcomes evolved.

The focus on design science has led to studies addressing the interaction of a variety of participants (e.g., students, teachers, researchers, curriculum developers), complex conceptual systems (e.g., complex programs of constructions, complex learning activities) and technology, all of which are influenced by certain social constraints and affordances (e.g., Schorr & Koellner-Clarke, 2003). Such studies contrast with previous research involving information-processing approaches that traced the cognitive growth of individuals in selected mathematical domains such as number (e.g., Simon & Klahr, 1995). In the intervening years, design science approaches have opened up a new world of mathematics education research—we now have a greater understanding of a "learning ecology" (Cobb et al., 2003). Mathematics education involves not just individual learners and teachers; rather, it involves complex, interacting systems of participants engaged in learning experiences of many types and at many levels of sophistication.

Design studies, however, have not been without their critics. Shavelson et al. (2003) warned that such studies, like all scientific research, must provide adequate warrants for the knowledge claims they make. As Shavelson et al. emphasized: "By their very nature, design studies are complex, multivariate, and interventionist, making warrants particularly difficult to establish" (p. 25). Furthermore, many of these studies rely on narrative accounts to relay and justify their findings; the veracity of such findings is not guaranteed. In addressing this concern, Shavelson et al., have provided a framework that links research questions evolving from design studies with corresponding research methods focused on validation of claims.

Included in their framework are considerations pertaining to: (a) "What is happening?" (e.g., characterizing a sample of students with a statistical sample, addressing the depth and breadth of a problem through survey, ethnographic, case study methods); (b) "Is there a systematic effect?" (addressing issues related to an intent to establish cause and effect); and (c) "Why or how is it happening?" (a question seeking a causal agent). In our efforts to move our discipline forward, we need to give serious consideration to concerns such as these, but at the same time, we need to ensure that design studies are not sidelined by policymakers' request for stringent, scientifically controlled research. We cannot afford to lose the rich insights gained from design research studies in our discipline.

INCREASED SOPHISTICATION AND AVAILABILITY OF TECHNOLOGY

Since the publication of my 1995 book, *Mathematics education: Models and Processes*, the increase in sophistication and availability of new technologies has been quite incredible. Such technological growth will escalate in years to come, making it difficult to predict what mathematics knowledge and understandings our students will need in even five years from now. What appears of increasing importance, however, is the need for students to solve a variety of unanticipated problems, to be innovative and adaptive in their dealings with the world, to communicate clearly their ideas and understandings in a variety of formats to a variety of audiences, and to understand and appreciate the viewpoints of others, both within the classroom context and globally.

Numerous opportunities are now available for both students and teachers to engage in mathematical experiences within international learning communities linked via videoconferencing and other computer networking facilities (e.g., see O'Neil & Perez, 2006). There are also increasing opportunities afforded by classroom connectivity where multiple devise types enable numerous representations to be passed bi-directionally and flexibly among students and between students and the teacher within the classroom environment (Kaput, Hegedus, & Lesh, in press). Kaput et al. view classroom connectivity as "a critical means to unleash the long unrealized potential of computational media in education."

As many researchers have emphasized, however, (e.g., Mauer, 2000; Moreno-Armella & Santos-Trigo, in press; Niss, 1999), the effective use of new technologies does not happen automatically and will not replace mathematics itself. Nor will technology lead to improvements in mathematical learning without improvements being made to the curriculum itself.

As students and teachers become more adept at capitalizing on technological opportunities, the more they need to understand, reflect on, and critically analyze their actions; and the more researchers need to address the impact of these technologies on students' and teachers' mathematical development (Niss, 1999).

CONCLUDING POINTS

I hope to have shown in this preface that the field of mathematics education has come quite some distance since the 1995 publication of *Mathematics education: Models and processes*. However, I have only touched upon some of the key developments that have shaped and continue to shape mathematics education as we know it today. We still need significantly more progress in our discipline. We need to find more effective ways of involving *all* students in meaningful mathematical learning—learning that will equip all students for a rapidly advancing and exciting technological world. But equally importantly, we need to ensure that all of our students have the mathematical competencies that will enable them to navigate successively through their daily lives and achieve the productive outcomes they desire. Mathematics transcends so many aspects of our existence; it is our role to ensure our students make maximum use of their mathematical achievements.

REFERENCES

Brownell, W. A. (1945). When is arithmetic meaningful? *Journal of Educational Research, 38*(3), 481-498.

Bruner, J. S. (1960). *The process of education.* Cambridge, MA: Harvard University Press.

Cobb, P., Confrey, J., diSessa, A., Lehrer, R., & Schauble, L. (2003). Design experiments in educational research. *Educational Researcher, 32*(1), 9-13.

Dienes, Z. (1960). *Building up mathematics.* London: Hutchinson Education.

English, L. D. (In press). Integrating complex systems within the mathematics curriculum. *Teaching Children Mathematics.*

English, L. D., & Halford, G. S. (1995). *Mathematics education: Models and processes.* Hillsdale, NJ: Lawrence Erlbaum.

Gainsburg, J. (2006). The mathematical modeling of structural engineers. *Mathematical Thinking and Learning, 8*(1), 3-36.

Goldin, G. (In press). Perspectives on representation in mathematical learning and problem solving. In L. D. English (Ed.), *Handbook of international research in mathematics education* (2nd ed.). NY: Routledge.

Gutstein, E. (2007). Connecting community, critical, and classical knowledge in teaching mathematics for social justice. In B. Sriraman (Ed.), *International perspectives on social justice in mathematics education* (pp. 109-118). Monograph 1of The Montana Enthusiast. Montana: The University of Montana and The Montana Council of Teachers of Mathematics.

Kaput, J., Hegedus, S., & Lesh, R. (2007). Technology becoming infrastructural in mathematics education. In R. Lesh, E. Hamilton, & J. Kaput (Eds.), *Models & Modeling as Foundations for the Future in Mathematics Education.* Mahwah, NJ: Lawrence Erlbaum.

Hjalmarson, M. A., & Lesh, R. (In press). Design Research: Engineering, Systems, Products, and Processes for Innovation. In L. D. English (Ed.), *Handbook of international research in mathematics education* (2nd ed.). NY: Routledge.

Hoyles, C., Noss, R., & Pozzi, S. (2001). Proportional reasoning in nursing practice. *Journal for Research in Mathematics Education, 32*(1), 4-27.

Hoyles, C., Bakker, A., Kent, P., & Noss, R. (In press). Attributing meanings to representations of data: The case of statistical process control. *Mathematical Thinking and Learning, 9*(4).

Lesh, R. (2006). Modeling students modeling abilities: The teaching and learning of complex systems in education. *The Journal of the Learning Sciences, 15* (1), 45-52.

Lesh, R., & Clarke, D. (2000). Formulating operational definitions of desired outcomes of instruction in mathematics and science education. In A. E. Kelly & R. A. Lesh (Eds.), *Handbook of research design in mathematics and science education* (pp. 113-149). Mahwah, NJ: Lawrence Erlbaum.

Lesh, R., & Doerr, H. M. (Eds.). (2003). *Beyond constructivism: Models and modeling perspectives on mathematic problem solving, learning and teaching.* Mahwah, NJ: Lawrence Erlbaum.

Lesh, R. & English, L. D. (2005). Trends in the evolution of models & modeling perspectives on mathematical learning and problem solving. *ZDM: The International Journal on Mathematics Education, 37* (6), 487-489.

Lesh, R. & Sriraman, B. (2005). Mathematics education as a design science. *ZDM: The International Journal on Mathematics Education, 37* (6), 490-505.

Lesh, R., & Zawojewski, J. S. (2007). Problem solving and modeling. In F. Lester (Ed.), *Second Handbook of research on mathematics teaching and learning.* Greenwich, CT: Information Age Publishing.

Moreno-Armella, L., & Santos-Trigo, M. (In press). Democratic access to powerful mathematics in a developing country. In L. D. English (Ed.), *Handbook of international research in mathematics education* (2nd ed.). NY: Routledge.

National Council of Teachers of Mathematics (2000). *Principles and standards for school mathematics.* Reston, VA: NCTM.

Niss, M. (1999). Aspects of the nature and state of research in mathematics education. *Educational Studies in Mathematics, 40,* 1-24.

O'Neill, H., & Perez, R. (Eds.). (2006). *Web-based learning: Theory, research and practice.* Mahwah, NJ: Lawrence Erlbaum.

Schorr, R., & Koellner-Clarke, K. (2003). Using a modeling approach to analyze the ways in which teachers consider new ways to teach mathematics. *Mathematical Thinking and Learning, 5*(2,3), 109-130.

Shavelson, R. J., Phillips, D. C., Towne, L., & Feuer, M. (2003). On the science of education design studies. *Educational Researcher, 32*(1), 25-28.

Simon, T. J., & Klahr, D. (1995). A computational theory of children's learning about number conservation. In T. J. Simon, & G. S. Halfrod (Eds.), Developing cognitive competence: New approaches to process modeling. Mahwah, NJ: Lawrence Erlbaum.

Van Engen, H. (1949). An analysis of meaning in arithmetic. Elementary School Journal, 49, 321-329, 395-400.

Wertheimer, M. (1959). *Productive thinking.* NY: Harper & Rowe.

CHAPTER 7

THE IMPACT OF ZOLTAN DIENES ON MATHEMATICS TEACHING IN THE UNITED STATES

James Hirstein
The University of Montana

I first encountered the writings of Zoltan Dienes as a young teacher of mathematics in a teacher education program. I had been prepared to be a teacher of mathematics in secondary school, but I was asked to teach some of the mathematics courses that would prepare teachers for elementary school. None of my preparation had introduced me to Dienes (I never saw anything by Bruner or Piaget either), and the prospective elementary teachers were asking questions about how children learn mathematics. This required attention to the structure of mathematics and to the psychology of learning mathematics. Dienes books and descriptions of lessons provided many of the answers I needed.

I first read *Building Up Mathematics* (Dienes, 1971) when it was in its fourth edition (it was originally published in 1960). Here he described his theory of six stages of learning mathematics: (1) free play, (2) games, (3)

Mathematics Education and the Legacy of Zoltan Paul Dienes, pages 107–111
Copyright © 2008 by Information Age Publishing
107

search for communalities, (4) representation, (5) symbolization, and (6) formalization (p. 36). I also read *The Power of Mathematics* (Dienes, 1964), where he amplified his theories on structure and representation. His examples taught me several early lessons.

First, the role of play and games is crucial in formulating the first understanding of a new concept. Students can be introduced to very complicated ideas and can develop quite sophisticated approaches to problems if things are presented at the right level. Second, abstraction and generalization are important skills that must be practiced. This means finding several "embodiments" of mathematical concepts that children can explore. And third, a consistent Dienes claim, symbolism usually occurs too soon. When children are required to use symbols before they understand what the symbols represent, the learning involves mostly memory and is not very long lasting.

Several of Dienes' inventions became standard equipment in the mathematics laboratory. His Multibase Arithmetic Blocks gave a concrete representation for number bases (see Figure 7.1 for a picture of the base 4 set). The principles of the base ten numeration system were so taken for granted that most students did not grasp the value of a base system. Dienes' Blocks allowed students to explore the numeration system and how the operations on numbers are addressed by the system. Algorithms for addition, subtraction, multiplication and division can be illustrated and explained in detail, whether using the standard versions or alternative versions. Students were free to invent their own algorithms, and what's more, they understood what they were doing.

A second set of materials from Dienes' fertile mind was the box of Logic Blocks (see *Learning Logic and Logical Games*, Dienes, 1974). The Logic Blocks were a completely balanced set of wooden pieces that varied in shape, color, size, and thickness. Every possible combination of the four attributes are available in the box. Examples of sets and their properties are easily shown; such as red pieces, non-red pieces, circle pieces, pieces that are red or circles, and so on. But just as important was the ability to define

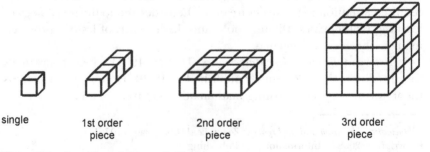

| single | 1st order piece | 2nd order piece | 3rd order piece |

Figure 7.1 Base Four Arithmetic Blocks

relationships between pieces; such as pieces that differ in one attribute, that differ in two attributes, that are the same shape, etc. Two pieces can be compared by "How are they alike?" or "How are they different?" Also, students can explore logical consequences in well-defined ways: "If a piece from this set is not a square, then it must be blue." Students who experienced activities using Dienes' materials often commented that the concrete examples made difficult concepts far more understandable. They supported Dienes' contention that trying to deal with mathematical symbolism (set notation) before the concepts were clear made learning difficult.

Not all of Dienes' examples are physical embodiments. One particular diagram (see Figure 7.2) is useful for several investigations of the additive structure or the multiplicative structure of number. For example, the horizontal (solid) arrow can represent "times 2" and the vertical (dashed) arrow can represent "times 3." With the leading circle set to 1, filling in the chart reveals many patterns of numbers, factors, and multiples. One can even go "backwards" along the arrows to see the relationships between fractions and inverses.

Dienes also gives many geometric illustrations. In a paper presented at a Symposium on Mathematics Laboratories (Dienes, 1975), he gives several concrete examples of finite geometries. As before, he gives multiple embodiments of the same structure and multiple representations that can be used to study geometric properties, such as duality and extensions. In his summary, he recapitulates his theory of the stages in this process:

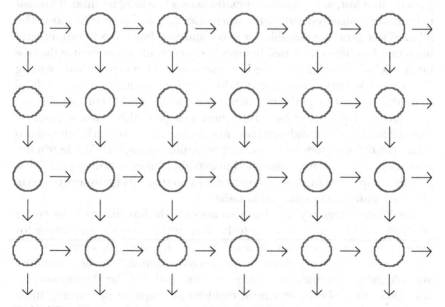

Figure 7.2.   Diagram for Generating Elements (Dimension 2)

> In the abstraction process that leads to the eventual formation of a formal-
> ized concept, there are many stages. The first is always a somewhat groping
> stage, a kind of "trial and error" activity; this is usually described as *play*. The
> restrictions in the play lead to rule-bound play or *games*. This has been well
> represented in the present paper. The next stage is the identification of many
> different games possessing the same structure. This is the stage of the search
> for *isomorphisms*. When the irrelevant features of the many games have been
> discarded, we are ready for a *representation*. Such are the many "link" diagrams
> suggested. It is only when this stage has been reached that it is fruitful to use a
> fully symbolic language, the development of which will be a later stage in the
> abstraction process. (pp. 83–84)

Have Dienes' ideas had an impact on school mathematics in the United States? At least indirectly, the influence has been major. Mathematics
laboratories have been a part of many mathematics classrooms since the
mid-1970s, and hundreds of the activities were designed or inspired by
Dienes' work. But in direct terms, one cannot be this positive. Most teachers would not recognize the contributions Dienes has made. Base 10 blocks
may be common, but other bases are not available. So many versions of the
Logic Blocks have been produced that Dienes rarely gets credit for these
activities. Geometry lessons today deal more with physical descriptions and
ignore logical arguments with abstract systems.

Part of the problem is due to timing and the cyclical nature of educational trends. Through the decade of the 1960s, the "new math" movement had
gained, then lost, acceptance as a mathematics teaching method. This new
math had emphasized abstract mathematics and formal justification. The
method was generally suitable for good students, but it was found difficult
for general audiences. Dienes theories were becoming known just as the new
math was losing popularity. The U.S. mathematics curriculum was heading
into a "back to basics" movement. Educators were concerned about school
students' abilities to perform calculations quickly and accurately. Dienes
was describing processes like abstraction and generalization and justification (all are clearly considered basic to a mathematicians' work, but general
educators did not understand his mathematical examples). While teacher
education programs may have tried to explain Dienes' learning principles,
only the simplest examples were meaningful to students (prospective teachers) with limited mathematical knowledge.

But Dienes' message has been consistent. He has managed to create
teaching activities that conform to his learning principles. His search for
embodiments of mathematical ideas has produced clever examples in a
wide variety of contexts. Music, motion, physics, dance, language, and even
abstract games—they all conform to the similarities of the abstraction process. Dienes has shown that most children are capable of learning these
sophisticated processes. The progression from concrete, through other

representations, to symbols and formal structures applies to all areas of knowledge. Dienes' great contribution has been that he has provided evidence of his principles at all levels and his activities use mathematical concepts. In many respects, it is best to demonstrate these processes in mathematics, so Dienes' examples will continue to inspire mathematics teachers and cognitive scientists for years to come.

REFERENCES

Dienes, Z. P. (1971). *Building Up Mathematics*. (4th ed.). London: Hutchinson.

Dienes, Z. P. (1964). *The Power of Mathematics*. London: Hutchinson.

Dienes, Z. P. (1974). *Learning Logic and Logical Games*. London: Hutchinson.

Dienes, Z. P. (1975). Abstraction and Generalization: Examples Using Finite Geometries. In J. Higgins (Ed.) *Cognitive Psychology and the Mathematics Laboratory*. Columbus, OH: ERIC/SMEAC.